EXCEL® STUDENT LABORATORY MANUAL AND WORKBOOK

JOHANNA HALSEY
Dutchess Community College

ELLENA REDA
Dutchess Community College

to accompany

THE TRIOLA STATISTICS SERIES:

Elementary Statistics, Tenth Edition

Elementary Statistics Using Excel, Third Edition

Essentials of Statistics, Third Edition

Elementary Statistics Using
the Graphing Calculator, Second Edition

Mario F. Triola
Dutchess Community College

PEARSON
Addison
Wesley

Boston San Francisco New York
London Toronto Sydney Tokyo Singapore Madrid
Mexico City Munich Paris Cape Town Hong Kong Montreal

Reproduced by Pearson Addison-Wesley from electronic files supplied by the authors.

Copyright © 2007 Pearson Education, Inc.
Publishing as Pearson Addison-Wesley, 75 Arlington Street, Boston, MA 02116

ISBN   0-321-36909-2

4 5 6 BB 09 08

PEARSON

Addison
Wesley

Preface

Purpose of the Manual

This manual contains step-by-step instructions to help you familiarize yourself with the spreadsheet program Excel. Our primary purpose in creating these instructions is to help you become proficient with those Excel features that support working with data in your statistics class.

We know it is impossible to cover every possible option in using the program, so we have chosen the techniques that have worked well for us and our own students. There are usually at least two ways to produce the same results. Our hope is that providing you with a solid set of step by step instructions, you will become comfortable enough with the program to begin to experiment on your own, and share your discoveries with other students in your class.

Layout of the Manual

Other than Chapter 1, this manual follows *Elementary Statistics, 10*[th] Edition section by section. The exercises worked through in the tutorial instructions are based on the presentation of the material in the corresponding section in your text. At the end of each section you will find reference to several exercises which will give you an opportunity to practice the technology skills introduced within that section. Our hope is that you will immediately employ the features you learn in Excel to help you with the exercises and projects presented in that section of your textbook. You will find that using the technology to work on many exercises in the book will afford you the practice necessary to become proficient with the program.

Using Technology Wisely

Any software package has its own learning curve. You should expect it to take a certain investment of time and energy and regular practice to become comfortable with using Excel. The more regularly you commit to using the ideas presented in this manual, the more proficient and adept you will become with using this program when and where appropriate.

While there are many, many places that Excel can, and should be integrated into the course material, you also can benefit from doing some of the work without using any technology. To truly understand some concepts, you need to perform at least some of the computations by hand. Once you have a core understanding of an idea, the technology affords you a way to find answers quickly and accurately.

Technology Notes

The instructions contained in this manual are written for a PC. Keystrokes may vary for those using a Mac. Mac users should note that Ctrl + click has the same functionality as a right click for PC users and a single click has the same functionality as a left click.

Early in the manual, you will be asked to load the Data Desk®/XL (DDXL) Add-In that accompanies your textbook. This Add-In supplements Excel, providing additional statistical tools not included in Excel. For example you will use DDXL to construct boxplots, confidence intervals and to perform hypothesis tests.

You can find the data sets from Appendix B of your textbook on the CD-ROM that accompanies your text. You can also download these data sets from the Internet at http://www.aw-bc.com/triola.

Final Notes

We feel the benefits of using Excel in a statistics course are vast. Many companies look for employees who are proficient with using spreadsheets. By learning how you can use Excel to support your work in statistics, you will simultaneously be developing a skill that is highly valued in the business world. Our hope is that this manual provides you with a relatively painless entry into the world of spreadsheets! We hope you enjoy your learning journey.

Johanna Halsey

Ellena Reda

CONTENTS

CHAPTER 1: GETTING STARTED WITH MICROSOFT EXCEL 1

CHAPTER 2: SUMMARIZING AND GRAPHING DATA 19

CHAPTER 3: DESCRIBING, EXPLORING, AND COMPARING DATA 37

CHAPTER 4: PROBABILITY 54

CHAPTER 5: DISCRETE PROBABILITY DISTRIBUTIONS 64

CHAPTER 6: NORMAL PROBABILITY DISTRIBUTIONS 76

CHAPTER 7: ESTIMATES AND SAMPLE SIZES 96

CHAPTER 8: HYPOTHESIS TESTING 105

CHAPTER 9: INFERENCES FROM TWO SAMPLES 111

CHAPTER 10: CORRELATION AND REGRESSION 123

CHAPTER 11: MULTINOMIAL EXPERIMENTS AND

CONTINGENCY TABLES 137

CHAPTER 12: ANALYSIS OF VARIANCE 141

CHAPTER 13: NONPARAMETRIC STATISTICS 146

CHAPTER 14: STATISTICAL PROCESS CONTROL 160

CHAPTER 1: GETTING STARTED WITH MICROSOFT EXCEL

SECTION 1-1: INTRODUCTION... 2

SECTION 1-2: THE BASICS.. 2

SECTION 1-3: ENTERING AND EDITING DATA INTO EXCEL.. 4

THE FORMULA BAR... 4

ENTERING TEXT AND NUMBERS 4

FORMATTING CELLS 5

SAVING YOUR WORK 6

EDITING INFORMATION 6

SELECT A RANGE OF CELLS 7

DROPPING AND DRAGGING 7

OPENING FILES IN EXCEL 8

SECTION 1-4: UNDERSTANDING AND USING FORMULAS:...................................... 9

ENTERING A FORMULA 10

COPYING A FORMULA 11

SECTION 1- 5: RELATIVE AND ABSOLUTE REFERENCE... 12

SECTION 1-6: MODIFYING YOUR WORKBOOK 12

INSERTING AND DELETING ROWS/COLUMNS.. 12

CHANGING COLUMN WIDTH AND ROW HEIGHT 13

ADDING WORKSHEETS TO A WORKBOOK.. 13

RENAMING A WORKSHEET 14

INSERTING COMMENTS 14

SECTION 1- 7: PRINTING YOUR EXCEL WORK... 15

SECTION 1- 8: GETTING HELP WHILE USING EXCEL... 17

TO PRACTICE THESE SKILLS 18

SECTION 1-1: INTRODUCTION

One of the most valuable computer programs used in business today is the spreadsheet. A spreadsheet program is an electronic workbook that allows you to organize and analyze data, to perform calculations and to show relationships in data through various types of charts and graphs. There are many practical applications for spreadsheet uses ranging from home to office use. Many computers today come preloaded with Microsoft Office®. Excel is part of the Microsoft Office Suite. It is a popular spreadsheet program and can be used by all skill levels. Excel has the capability to perform tedious operations on very large sets of data quickly.

The purpose of this chapter is to provide you, the student, with an introduction to spreadsheets and to prepare you to use Excel in the study of statistics. If you have not had any experience with a spreadsheet program this chapter will be invaluable to you. If you have had limited exposure to Excel this chapter will provide the review you will need to be proficient with this program as you utilize it throughout your statistics course.

The best way to learn the basics is to dive in. As you explore Excel you will notice that there are often several ways to perform the same task. This manual will highlight only one or two of those ways. However that should not prevent you from trying other ways or using a method you already are familiar with.

SECTION 1-2: THE BASICS

You may open the Excel program one of two ways:

1) **Double click on the Excel icon** found on your desktop.

    OR

2) From the **Start** menu

    a. Highlight **Programs**

    b. Highlight **Microsoft Excel**

    c. **Click on the left mouse button** (Mac users just click).

When you start Excel, a blank worksheet appears. This worksheet is the document that Excel uses for storing and manipulating data.

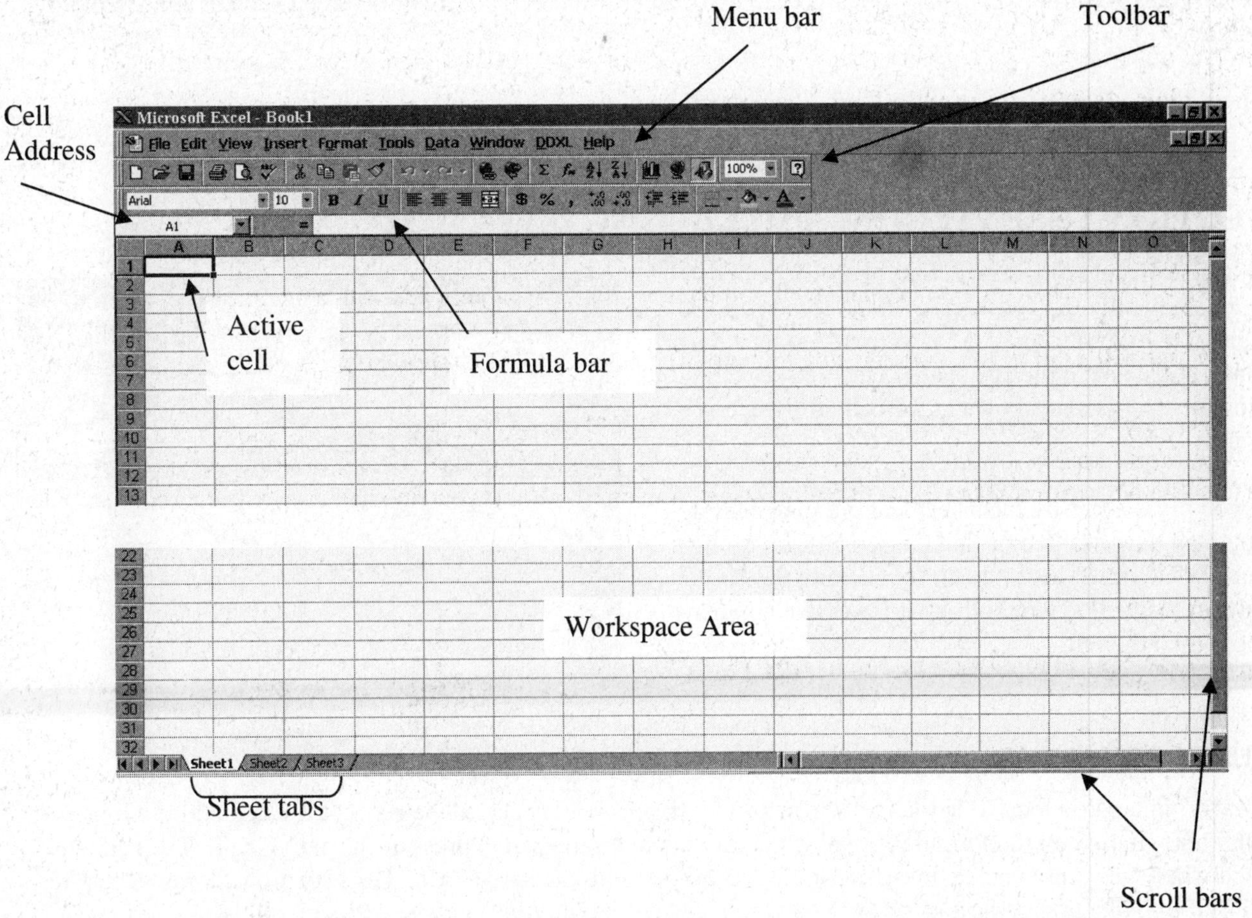

Menu bar   Toolbar

Cell
Address

Active
cell

Formula bar

Workspace Area

Sheet tabs

Scroll bars

The following are some **basic terms** that you should be familiar with. Locate each of these on your Excel screen:

Toolbar – An area of the Excel screen which contains a variety of icon buttons that are used to access commands and other features. To find out what each toolbar icon does just place the mouse pointer over the button without clicking and its name will appear.

Menu bar – Groups of command choices. To view those choices click on one of the commands and a menu of additional commands for that group will drop down.

Worksheet area – The grid of rows and columns into which you enter text, numbers and formulas.

Cell – Located at the intersection of a row and a column. Information is inserted into a cell by clicking on cell and entering the information directly.

Cell address – Location of a cell based on the intersection of a row and a column. In a cell address the column is always listed first and the row second so that A1 means column A row 1.

Active cell – The worksheet cell receiving the information you type. The active cell is surrounded by a thick border. The address of the active cell is displayed above the worksheet on the left.

Formula bar – Area near the top of the Excel screen where you enter and edit data.

Scroll bars – Allow you to display parts of the worksheet that are currently off screen such as row 35 or column R.

Sheet tabs – Identify the names of individual worksheets.

SECTION 1-3: ENTERING AND EDITING DATA INTO EXCEL

When an Excel worksheet is first opened, the cell A1 is automatically the active cell. In the screen on the right you will notice that A1 is surrounded by a dark black box. This indicates that it is the **active cell**. Note that A1 is also shown in the **cell address** box as well.

If you press a key by mistake, clear out information you may have wanted or find the screen isn't doing quite what you expect it to do try clicking the **Undo button** located on the toolbar. It looks like an arrow looping to the left.

THE FORMULA BAR

The formula bar is located in the fourth row of your Excel worksheet, to the right of the cell address box. The first window on the formula bar shown below indicates the cell address of the active cell. The **red X** is used when we wish to delete information we have typed into the active cell. The **green check mark** can be clicked to indicate that the data or formula you have entered into the active cell is acceptable.

There are three types of information that may be entered into a cell:

- Text
- Numbers
- Formulas

ENTERING TEXT AND NUMBERS

The following exercise will give you some practice in entering both text and numbers into a worksheet.

1) Position the mouse pointer on cell A1 and click on the left mouse button. This makes cell A1 the active cell.

2) Type MONTHLY EXPENSES in cell A1 and press **Enter.** Notice that the active cell is now A2.

3) In cells A2 through A7 type Rent, Food, Utilities, Cable, Telephone and Internet. Press the **Enter** key or the **down arrow** key to move down to the next cell.

4) Similarly you will enter the amount spent each month in these categories in cells B2 through B7. To complete this activity begin by activating cell B2 and typing in the data as shown below:

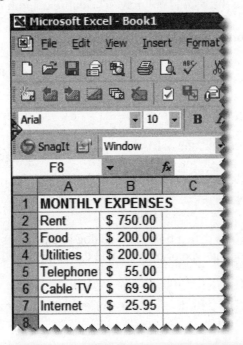

Note:
To move to the cell just below a cell you have entered data in press **Enter.**
To move to the right of a cell you have just entered data in press **Tab.**

FORMATTING CELLS

The information that you just entered into the Excel spreadsheet represents dollar amounts, although this may not be apparent when you first enter the data.

The following shows the formatting icons found in your Excel worksheet.

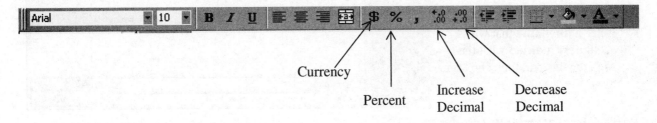

Currency

Percent

Increase Decimal

Decrease Decimal

The **increase decimal or decrease decimal** icons both allow us to determine the number of values that will be displayed to the right of the **decimal.**

To Format Cells

There are two different ways to change the values you have entered so that they will represent dollar amounts. These are outlined on the following page.

Method 1:

1) **Click on** the **B** at the top of the second column. The entire column is now highlighted.

2) **Click on the currency icon** ($) found on the formatting tool bar.

3) All data in that column should now contain a dollar sign in front of it.

Method 2:

1) **Select the cells** you wish to format.

2) **Click on Format** from the Menu Bar.

3) **Highlight Cells** and click.

4) This will open a **Format Cells** dialog box. This dialog box presents a number of options to you, that include the ability to choose the type of number and the number of decimal places you wish displayed.

Try using this method to change all of your data in your worksheet to dollar amounts at the same time.

SAVING YOUR WORK

Once you have entered data into an Excel spreadsheet you will want to save your work. This can be done by either clicking the **Save** icon on the toolbar (it looks like a floppy disk) or by clicking on **File** in the menu bar and choosing the **Save** command.

- A dialog box will appear similar to the one you see on the right.

- Choose **Drive A** to save the information to a **floppy disk**.

- Choose **C** to save information to your **hard drive.**

- Enter a **file name** that will accurately indicate what the file contains and then click on **Save**.

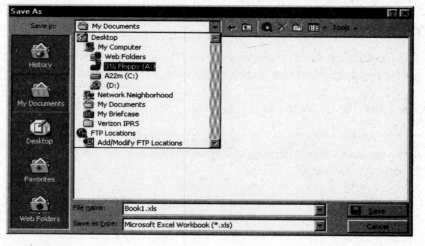

Note:
It is a good idea to follow the wise old adage of "save and save often". It can be very frustrating to spend a fair amount of time working on something only to lose your work due to a computer glitch. Be mindful that when working on a spreadsheet, it is always a good idea to click on the **Save** icon periodically.

EDITING INFORMATION

After data has been entered you can edit the information in a cell by activating that cell (click on it). The information in that cell is now displayed in the formula bar. Move your cursor to the entry you wish to modify or change. The cursor will turn into what looks like the capital letter I, commonly referred to as an I-beam. Place the I-beam at the point you wish to make changes, left click and proceed from there.

SELECT A RANGE OF CELLS

We often need to select more than one cell at a time. A group of selected cells is called a **Range** of cells.

To select more than one cell at a time

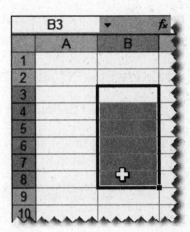

1) **Click on a cell** (in this case we activated cell B3) and hold the left mouse button down.

2) With the **cursor in the middle of the cell** move down to cell B8.

3) **Release the left mouse button.** The cells B3 through B8 should now be highlighted.

We refer to this **range of cells** as **B3:B8**, using a colon to separate the first cell in the range from the last cell.

DROPPING AND DRAGGING

To easily move or copy a range of cells from your worksheet we can use a shortcut method known as **drop and drag**.

To move a range of cells

1) Select the cells using the method previously outlined

2) Position the cursor anywhere on the border of the range of cells you have highlighted.

3) Hold the left mouse key down.

4) Drag the range of cells to their new location.

5) Release the mouse button.

To copy a range of cells

1) Select the cells as you did above, release the mouse

2) Position the cursor anywhere on the border of the range of cells you have highlighted.

3) Hold the left mouse key and the **CTRL** key down at the same time.

4) Move the data to the location into which you wish it to be copied.

5) Release the mouse key, then the **CTRL** key. (The order in which you release the keys is important).

Note:
You can **Undo Drop and Drag** by locating this command in the **Edit** menu.

OPENING FILES IN EXCEL

As you work through this manual and on exercises in your textbook, you will be asked to open files you have saved to a floppy disk or to your hard drive. You will also be asked to use data sets found on the CD-ROM Data Disk that accompanied your Statistics textbook. These data sets are the same as the data found in Appendix B of your text.

To use the data sets on the CD-ROM

1) Open Excel.

2) **Place the CD into the CD-ROM** drive on your computer. This is often drive D or E.

3) **Click on File**, found on the menu bar, **highlight Open** and click.

4) **A dialog box similar** to the one shown below. For information stored on the CD choose **D** if this is the **CD – Rom drive** for your computer.

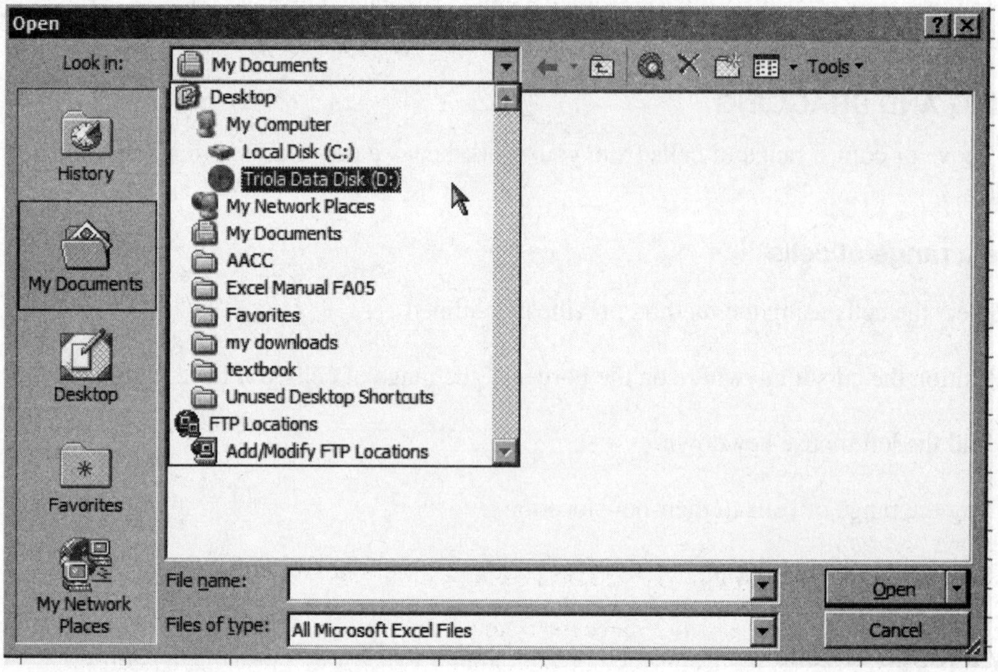

5) A second dialog box opens. This dialog box shows two files "DataSets" and "Software". Double click on **DataSets**.

6) This presents another dialog box which is shown on the following page. In this dialog box choose the **Excel folder** for a complete list of the files for Excel. This can be seen on the following page as well. Use the scroll bar on the right of the dialog box to view the complete list of data files.

7) Locate the file you are looking for by file name. Click on the **file name**.

8) Click on **Open.**

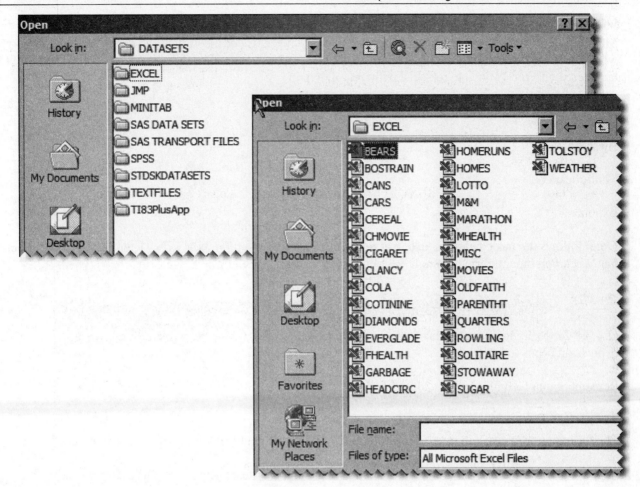

The same process is used when opening a file that contains data that you may save saved to a floppy disk, usually drive A, or to your hard drive, usually drive C. .

SECTION 1-4: UNDERSTANDING AND USING FORMULAS:

When we analyze data in Excel it is often necessary to perform calculations using formulas. When we enter a formula in Excel we can either use the actual value found within a cell or the cell address.

1) When you type formulas in Excel it is important that you **place an equal sign (=) at the start of your formula.** This is how you communicate to the software that you are entering a formula rather than data or text.

2) As you **type your formula** it will be displayed in the formula bar.

3) After you have typed in the formula, **press the Enter** key. Excel automatically performs the calculation and displays the numerical result in the cell you have chosen.

4) If you change a number stored in a cell address used in the formula, Excel automatically recalculates the results of that formula.

Operators & Order of Operation

Many of the mathematical formulas that we use in Excel make use of the same basic arithmetic operations with which we are all familiar. Some of the arithmetic symbols we use for these **common operators** need to be entered differently and are replaced with the following:

Addition	+		Subtraction	—
Multiplication	*		Division	/
Exponents	^			

Excel follows the basic **rules for order of operation** that we are all familiar with. Therefore it is important that when you input a formula into Excel you are very specific in the way the information is typed.

$\dfrac{2 + 3(2)^5}{5}$ would be entered into Excel as follows: (2 + 3 * 2 ^ 5)/5. The parentheses around the expression (2 + 3 * 2 ^ 5) are important because they indicate that (2 + 3 * 2 ^ 5) is the numerator and the entire expression needs to be divided by 5.

Remember that formulas and equations generally contain a cell address rather than actual numbers.

Entering a Formula

The **following exercise** will give you some practice in **entering a formula** into an Excel worksheet.

Suppose you wish to use Excel to determine the average of your test scores for a particular course.

1) In cell A1 type the heading SCORES

2) In cells A2 through A7 enter the following test grades: 89, 73, 95, 85, 69 and 100.

3) In cell C2 type the word AVERAGE

4) **Select a cell** to the right of C2 or below C2 **to enter the formula** for finding the average of these scores.

5) **Type an = (equal sign) followed by the formula**. The formula should be surrounded by a set of parentheses. It is often easier to use the cell address within any formulas you write rather than the actual number although you can use either one. Using a cell address is advantageous when copying a formula to other cells.

6) **Type** =(A2+A3+A4+A5+A6+A7)/6

7) After typing in the formula **press Enter**.

8) You may need to use the **decrease decimal** icon to round your result off to the nearest whole number.

Notice that C3 has a black border around it indicating that it is the active cell. Also notice that the formula used to find the average in C3 is displayed in the formula bar.

Note:

Typing in each cell address to be used in finding the average works well if there is a small number of cell addresses to be entered. If we had 50 test scores in column A that we would not want to type = (A1 + A2 + A3 + …. + A50).

To save time we can enter the same information by typing in the cell address for range of cells, in this case type = AVERAGE(A1:A50).

Copying a Formula

Suppose your instructor kept all of her students' grades in an Excel worksheet and used the capabilities of this software program to determine each student's average at the end of the semester. Rather than retype the formula for finding the average grade for each individual student your instructor can simply copy an existing formula into other cells. The following outlines two methods that can be used to copy a formula from one cell to another.

Method 1: (useful if you are copying the same formula to a series of cells)

1) **Click on the cell** that already contains the formula you want to copy.

2) Place your mouse on the lower right hand corner of the highlighted cell. When your cursor changes to a cross (commonly referred to as the **fill handle**), click and hold the left mouse button and drag the box to cover the cells where you wish to copy the formula.

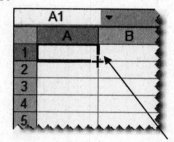

Fill handle

3) When you release the mouse button the formula will be copied and adjusted for these cells.

Method 2: (useful if you are only copying one piece of information to another cell)

1) Click on the cell that already contains the formula you want to copy.

2) Click on the copy icon found on the tool bar (or use Ctrl + C).

3) Click on the new cell into which you wish to copy the formula.

4) Choose the paste icon found on the toolbar. (Or use Ctrl + V).

Try both methods on the worksheet you have created to determine which works best for you.

SECTION 1-5: RELATIVE AND ABSOLUTE REFERENCE

Cell references are referred to as either **relative cell reference** or **absolute cell reference.** The cell references we have used so far are all **relative cell references** and when these references are copied to a new location they change to reflect their new position. This relative address feature makes it easy for us to copy a formula by entering it once in a cell and then copying its contents to other cells.

An **absolute reference** does not automatically adjust when moved to another cell and is used when it is necessary to retain the value in a specific cell address when copying a formula. In an absolute reference both values in the cell address are preceded by a $. For example, the formula = ($B$5 + $C$5 + $D$5 + $E$5) will remain unchanged regardless of the cell to which it is copied. The dollar sign does not signify currency but rather is used to identify that the cell is an absolute reference.

Formulas can contain both relative and absolute cell references. In Section 1 – 3 we entered data into an Excel worksheet outlining monthly expenses. Suppose we wish to determine what percentage each expense represents of the total expenses for the month.
-   We begin by totaling the expenses listed in column B. This information can be found in cell B9 as seen below.

-   To determine the percentage each individual expense represents of the total we divide the individual expense by the total expenses OR B2/$B$9. Enter this formula in cell D2.

The use of an absolute reference will allow us to copy this formula down column D. The top value will adjust to reference the cell we are in while the bottom value (the absolute reference) will remain constant, as we would want it to. You may need to format your percentages so that they are displayed as seen below.

	A	B	C	D	E
	L3		*fx*		
1	MONTHLY EXPENSES			Percent of Expenses	
2	Rent	$ 750.00		58%	
3	Food	$ 200.00		15%	
4	Utilities	$ 200.00		15%	
5	Telephone	$ 55.00		4%	
6	Cable TV	$ 69.90		5%	
7	Internet	$ 25.95		2%	
8					
9	Total	$1,300.85			
10					

SECTION 1-6: MODIFYING YOUR WORKBOOK

INSERTING AND DELETING ROWS/COLUMNS

Sometimes you will want to insert or delete columns or rows from your worksheet. If you want to create additional space in the middle of a worksheet you can insert a column or a row that will run the entire length or width of the worksheet. If you have an entire row or column that is no longer necessary you can delete the entire column or row.

To Insert a Row (or Column)

1) **Highlight the row** (or column) by right clicking on the number at the start of the row (or letter at the head of the column). A dialog box should open.

2) **Choose INSERT**. The new row will be inserted *above* the selected row. A new column will be inserted to the *left* of the selected column.

To Delete a Row (or Column)

1) **Highlight** the row (or column) by right clicking on the number at the start of the row (or letter at the head of the column). A dialog box should open.

2) **Choose DELETE**. The old row (or column) is removed from the worksheet and is replaced by the data in the adjoining row (or column).

Note:

Pressing the **Delete** key does not delete the selected row or column. It clears all data from the selection without moving in replacement data. Click a cell outside of the selected row (or column) to deselect it.

CHANGING COLUMN WIDTH AND ROW HEIGHT

Often we are in a position where we wish to change the width of a column. Text is often cut off because the column is not wide enough to display all that we have entered into the cell. If a cell can't display an entire number or date, the cell may fill with ####### or display a value in scientific notation. (Try entering a 12 digit number into a cell.)

To Adjust the Column Width

1) Position the mouse pointer on the right border of the lettered heading at the top of the column you wish to adjust. The mouse pointer should change to a black cross with an arrow head at each end of the horizontal line.

2) Press and hold down the left mouse button, dragging the right side of the column to increase or decrease the column width.

3) You can move the mouse back and forth. Release the mouse button when you have a column the width you like.

To Adjust the Row Height

1) Position the mouse pointer on the lower border of the numbered row whose size you wish to adjust. The mouse pointer should change to a black cross with an arrowhead at each end of the horizontal line.

2) Press and hold down the left mouse button, dragging the border to increase or decrease the row height.

3) Release the mouse button when you have a row at the height you like.

To Add Worksheets to a Workbook

When you open a new workbook in Excel you will see that there are three worksheets available to you. This gives you the ability to do three separate problems or variations of a problem all within one workbook. While only three worksheets are presented, it is possible to have up to sixteen worksheets within one workbook. To add a worksheet to your workbook:

1) **Click** on **Insert** on the menu Bar,

2) **Highlight Worksheet**, and then click.

You will notice an additional worksheet tab in the bottom area of the screen.

To Rename a Worksheet

Initially the worksheets in Excel are labeled "Sheet 1" "Sheet 2" and "Sheet 3". It is generally a good idea to rename your worksheets. This allows you to keep track of the information, data and graphs which are typically found in an Excel worksheet. To rename your worksheet:

1) **Right click on the worksheet tab** that you would like to rename

2) **Highlight Rename** and click. The sheet tab should now be highlighted.

3) Begin **typing the new name** for this worksheet.

4) Press **Enter** when you have finished.

INSERTING A COMMENT

There are many instances when it is important to annotate work done in Excel by attaching a note to a cell. This is done by adding comments that can be viewed when you rest the cursor over that cell.

To enter a comment into an Excel worksheet:

1) **Click the cell** to which you want to add the **comment**.

2) **Click** on the **Insert** menu.

3) **Highlight Comment**, the click.

4) Type your comment in the textbox that opens. When you finish typing the text, click outside the comment box.

	A	B	C	D	E
1	MONTHLY EXPENSES			Percent of Expenses	
2	Rent	$ 750.00		58%	
3	Food	$ 200.00		15%	
4	Utilities	$ 200.00		15%	
5	Telephone	$ 55.00		4%	
6	Cable TV	$ 69.90		5%	
7	Internet	$ 25.95		2%	
8					
9	Total	$1,300.85			
10					

5) A **red triangle** will appear in the upper right hand corner of the **cell that contains a comment**.

6) To view the comment, move the cursor over the cell that contains the note.

	A	B	C	D	E	F
1	MONTHLY EXPENSES			Percent of Ex		
2	Rent	$ 750.00		58%		
3	Food	$ 200.00		15%		
4	Utilities	$ 200.00		15%		
5	Telephone	$ 55.00		4%		
6	Cable TV	$ 69.90		5%		
7	Internet	$ 25.95		2%		
8						
9	Total	$1,300.85				
10						

E Reda
More than half of my monthly expenses goes for rent.

7) The comment will appear. This can be seen in the screen shot shown on the right.

To Show, Edit or Remove a Comment:

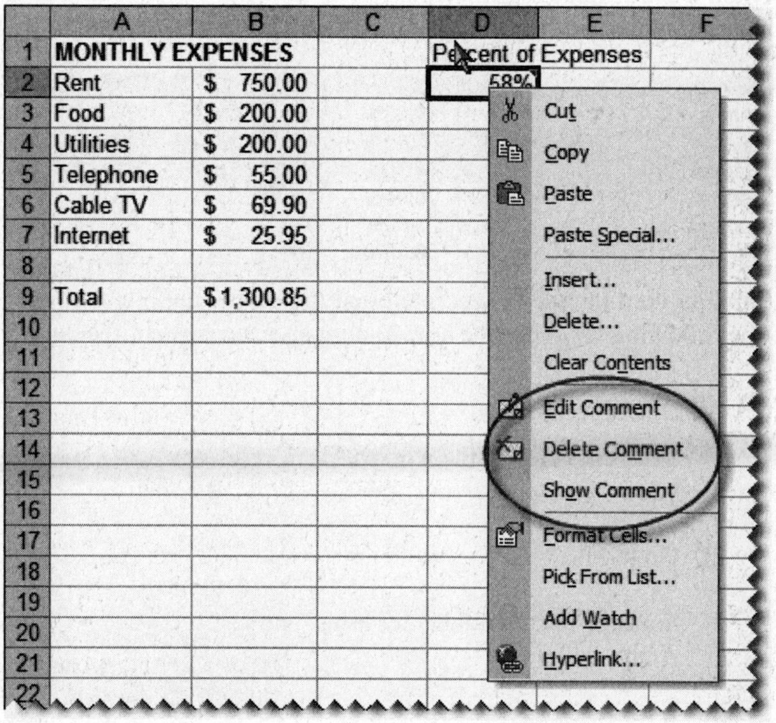

To make a comment **visible** at all times:

1) **Right click** on the cell that contains the comment.

2) Highlight **Show Comment,** and then click.

To **edit** a comment:

1) **Right click** on the cell that contains the comment.

2) Highlight **Edit Comment**, and then click.

To **remove** a comment:

1) **Right click** on the cell that contains the comment.

2) Highlight **Delete Comment,** and then click.

	A	B	C	D	E	F
1	MONTHLY EXPENSES				Percent of Expenses	
2	Rent	$ 750.00		58%		
3	Food	$ 200.00				
4	Utilities	$ 200.00				
5	Telephone	$ 55.00				
6	Cable TV	$ 69.90				
7	Internet	$ 25.95				
8						
9	Total	$ 1,300.85				

Cut
Copy
Paste
Paste Special...
Insert...
Delete...
Clear Contents
Edit Comment
Delete Comment
Show Comment
Format Cells...
Pick From List...
Add Watch
Hyperlink...

SECTION 1- 7: PRINTING YOUR EXCEL WORK

There are many occasions where you will want to print a worksheet or workbook that you have created in Excel. While it is possible to print all of the sheets in a workbook in one operation it is highly recommended that you print each worksheet separately.

PRINTING A WORKSHEET

Print Preview:

It is *always recommended* that you use the **Print Preview** before printing anything. This will give you an opportunity to make sure all of your work is presented within the printable page. It will also give you a chance to catch and correct mistakes before you print a page.

To preview your worksheet before you print it:

1. Click on **File** from the Menu Bar

2. Highlight **Print Preview** and click

3. Click on **Close** once you have previewed your worksheet.

Setting the Print Area:

Often we only need to print a portion of an Excel worksheet. For example. Suppose your teacher has entered her class grades into an Excel spreadsheet. A class roster might look like this:

Brandon	62.00%	15		51		39	53.5	38	52	23.0
Jennifer	89.00%	15	61.0	84	10.0	45.0	73.5	34	55	13
Travis	92.00%	15	57	92	10	45	72.0	50	58	13
Jannette	54.00%	15	52.5	51	10	38	33.0		23	21
Jessica	70.00%	15	34	70		35	66.5	39	56	34
Charlene	63.00%	13	23	66		25	55.5	19.5	50	27
Jing	88.00%	15	62	65		45	74.0	47	40	34

Suppose your instructor was interested in printing out just the first two columns of this Excel worksheet. The following steps outline how to print specific parts of the current worksheet.

To set the print area:

1) Select the area of the worksheet you wish to print.

2) Go to File

3) Select Print Area, then Set Print Area.

4) If this is done correctly you will see a dotted line around the area you have chosen to print.
When you print the spreadsheet, only the print area you have set will be printed.

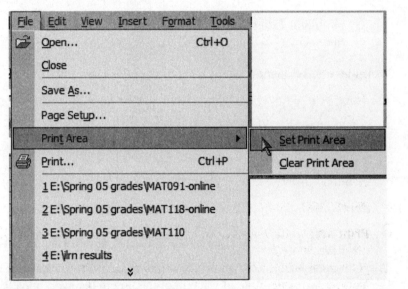

To Print:

1) Click on **File**, highlight **Print** and click.

2) **A Print dialog box** will appear and should be similar to the one shown on the right.

3) Check **Active sheet** to print your current worksheet.

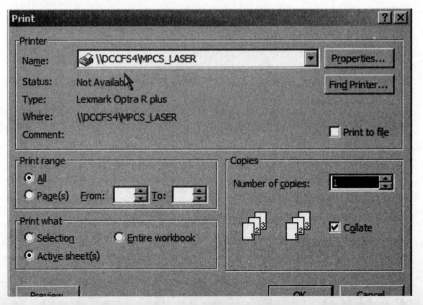

Check **Entire workbook** to print all of the worksheets in your current Excel workbook

4) You can also customize your worksheet before printing. Click on **File**, highlight **Page Setup** and click.

The following briefly describe some of the options available to you in **Page Setup.**

• **Page Tab** – allows you to choose the orientation of the printed page, either vertically (Portrait) or horizontally (Landscape). Landscape is useful for worksheets that have a greater width than they do length.

• **Margin Tab** – allows you to adjust the margin of the printed page.

• **Header/Footer Tab** – gives you the opportunity to include headers or footers on the printed page.

• **Sheet Tab** – the Print area is the most useful option on this tab.

• **Print area** - allows you to choose whether or not to include gridlines and row and column headings in the printout of your worksheet. It also gives you the opportunity to select a particular range to print from your worksheet. Click in the box to the right of Print area; highlight the range of cells you wish to print. You can also enter the information by typing in the range of cells.

Once you have completed making your choices click **OK**.

SECTION 1- 8: GETTING HELP WHILE USING EXCEL

Excel comes with a complete on-line Help feature designed to give assistance when you are having difficulty with a topic. You can access the Help system by selecting an option from the **HELP** Menu located at the top of the screen. Choose either Microsoft Excel Help or Contents and Index.

• **Microsoft Excel Help** gives you the opportunity to type in a question and search for an answer.

• **Contents and Index**:
  Contents is like the table of contents in a book providing an overview of major categories.

Index is like the index in the back of a book providing an alphabetical list of the help that is available.

TO PRACTICE THESE SKILLS

It is important to practice those technology skills introduced in this chapter before moving on. To help you do this work through the following problem in Excel.

Temperature Conversions

The formula for converting degrees Fahrenheit to degrees Celsius is $C = \dfrac{5}{9}(F - 32)$. Use Excel to set up a spread sheet to do these conversions given a set of temperatures.

a) Type "Degrees Fahrenheit" in Cell A1.

b) Type "Degrees Celsius" in Cell B1.

c) In cells A2 through A11 enter the following temperatures recorded in degrees Fahrenheit:
   $-10°, 0°, 10°, 32°, 45°, 50°, 68°, 75°, 83°, 95°$. (You do not have to type in the degree symbol.)

d) In cell B2 enter the formula to convert from degrees Fahrenheit to degrees Celsius using an appropriate cell address.

e) Copy this formula through to cell B11.

f) Format the values in column B correct to 2 decimal places.

g) Rename your worksheet "Temperature Conversion."

h) Print out your worksheet.

CHAPTER 2: SUMMARIZING AND GRAPHING DATA

SECTION 2-1: OVERVIEW & ADD-INS 20

    ADD-IN: LOADING ANALYSIS TOOLPAK 20

    ADD-IN: LOADING DDXL 20

SECTION 2-2: FREQUENCY DISTRIBUTIONS AND HISTOGRAMS FROM DATA 21

    OPENING AND ORGANIZING YOUR DATA 21

    CREATING AN INITIAL FREQUENCY DISTRIBUTION AND HISTOGRAM 22

    MODIFYING THE INITIAL FREQUENCY DISTRIBUTION 24

    MODIFYING THE INITIAL HISTOGRAM 26

    SELECTING REGIONS IN A HISTOGRAM & ACCESSING FORMATTING OPTIONS 27

    TO PRACTICE THESE SKILLS 28

SECTION 2-3: RELATIVE FREQUENCY HISTOGRAMS GIVEN A FREQUENCY DISTRIBUTION 28

    CREATING A RELATIVE FREQUENCY DISTRIBUTION 28

    CREATING A HISTOGRAM FROM A GIVEN FREQUENCY TABLE USING CHART WIZARD 30

    TO PRACTICE THESE SKILLS 31

SECTION 2-4: CREATING OTHER GRAPHS 32

    CREATING A FREQUENCY POLYGON 32

    CREATING A PARETO CHART 33

    CREATING A PIE CHART 35

    PRINTING A GRAPH 36

    TO PRACTICE THESE SKILLS 36

SECTION 2-1: OVERVIEW & ADD-INS

In this chapter, we will use the capabilities of Excel to help us summarize and graph sets of data. The sections that follow take you through step by step directions on how to create frequency distributions and histograms, as well as other types of visual models.

There is an Add-In that is necessary when working directly with Excel to create frequency tables and histograms called **Analysis ToolPak**. Depending on how you initially installed Excel, you may need to insert your original Excel software disk to access this Add-In. If you need the disk, you will be prompted to insert the disk into a drive as part of the process you work through below.

A second Add-In is available on the CD that comes with your text book, and is called **DDXL**. This Add-In makes it possible to do some work within Excel that may be easier to accomplish than just using Excel alone, and includes some features that are not available directly through the Excel program. We suggest you install both these Add-Ins right from the start, so they are available to you when you want to access them.

ADD-IN: LOADING ANALYSIS TOOLPAK

Before you begin your work in this chapter, make sure that you have the **Analysis ToolPak** loaded on your machine. This is an Add-In in Excel. You may need to access your Excel software disk to add this feature. If the disk is required, you will be prompted to insert it as you follow the steps below.

To see if the ToolPak has already been loaded on your machine, click on **Tools** in the menu bar, and see if **Data Analysis** is listed. If not, follow these steps:

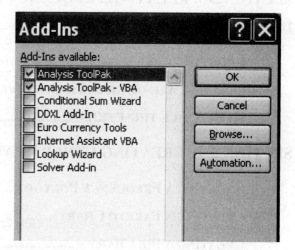

1) Click on **Tools**, and select **Add-Ins**.

2) In the Add-Ins dialog box, check Analysis Tool-Pak., as well as Analysis TookPak-VBA.

3) Click on **OK**. The ToolPak will be loaded, and when you click on **Tools** you should now see **Data Analysis** listed at the bottom of the menu.

ADD-IN: LOADING DDXL

DDXL is an add-in for Excel which is included on the CD included with your book. This add-in allows you to work with data which is presented in an Excel worksheet.

1) Press Start, and then select Run from the toolbar at the bottom of your Windows screen.

2) In the Run Dialog box, press Browse, and double click on the drive containing your CD. This drive should now show the name "Triola" indicating that the disk in the drive is the disk that accompanies your textbook.

3) Double click on the file folder that says Software.

4) One of the options in this folder is the DDXL program. Double clicking on this option will show 2 further options. Double click on the option that says "Install DDXL."

5) You should be taken back to your Run Dialog box, and should now see something similar to the following in the box: "E:\Software\DDXL\Install DDXL.exe." Press OK. You will be taken through the Install Wizard for the program. Follow the instructions on the screen. Make sure you

pay attention to where the program is being installed! You will need to know this to add the program into your Excel program. Typically the install process will add the program to your hard drive in your Program Files Folder.

6) Open Excel, and from the main menu bar, select Tools, Add-Ins. Then click on Browse.

7) Click on the pull down arrow at the end of the "Look In" box. Select the location where the DDXL program was installed. Again, typically you will select your C drive, and then select the Program Files Folder. Once there, you should see a folder called DDXL. Clicking on this opens up a screen which should show "DDXL Add-In." Select this option by either double clicking on it, or by clicking on it and then pressing OK.

8) You will be taken back to your Excel Add-In Selection Box, and you should see "DDXL Add-In" listed with a check in the box preceding the name. Click on OK.

9) You should now see DDXL listed in the main menu bar.

SECTION 2-2: FREQUENCY DISTRIBUTIONS AND HISTOGRAMS FROM DATA

In this section, we will learn how to create a frequency distribution and histogram directly from a data set. We will use the data from **Data Set 4 in Appendix B**. This data set shows measured levels of serum cotinine (a metabolite of nicotine) for smokers, nonsmokers exposed to environmental tobacco smoke, and nonsmokers with no environmental exposure to tobacco smoke.

OPENING AND ORGANIZING YOUR DATA

Developing Good Habits: You should get in the habit of preserving your original data in one of the sheets in your workbook, and then working with the specific data of interest in another worksheet. This way, you can always easily retrieve it if something goes awry as you work with specific parts of the original data. You should also clearly name the worksheet tabs so that it is clear what is contained in each sheet.

1) **Load the data into Excel:** You can access the data from the CD that came with your text book. The file name is COTININE.XLS. You can also find the data set at the website for the textbook: www.aw-bc.com/triola.

2) **Preserve your original data in a clearly marked worksheet:** After loading the original data from your CD, or from the website, double click on the tab at the bottom of the worksheet so that "Sheet 1" is highlighted. Type in "Original Data."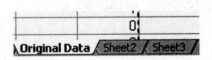

3) **Select the data you will be working with:** For our work, we will be working with the data which shows the measured levels of serum cotinine for Smokers. If you loaded the data from your CD, your data appears in columns A, B and C. Since the data we want to summarize is in column A, select this column by clicking on the "A". This column should now be highlighted.

4) **Copy the selected data to a new worksheet:**

   a. Once you have selected the data you are going to work with, copy this data by selecting **Edit** from your menu bar, and then selecting **Copy**. (**Alternatively, you can press Ctrl + C.**) You should see a dotted box around your selected data.

   b. At the bottom of your worksheet, click on Sheet 2. Position your cursor at the top of column A and click. This column should now be selected.

c. Click on **Edit** in your menu bar, and then select **Paste**. (**Alternatively, you can press Ctrl + V.**) Your data should be copied into Sheet 2.

d. **Rename Sheet 2:** Since we will be creating a frequency distribution and histogram for the data about Smokers, you may want to rename your "Sheet 2" with a name like "Smokers". Remember, to accomplish this, double click on the tab, and once it is highlighted, type in "Smokers".

CREATING AN INITIAL FREQUENCY DISTRIBUTION AND HISTOGRAM

In order to create a frequency distribution, we need to indicate what **upper class limits** we want to use. Excel will refer to these upper class limits as "bins."

Before using Excel, you need to determine how many classes you want, what your class width will be, and what your lower and upper class limits will be for the particular data set. The steps are outlined in your book under "Procedure for Constructing a Frequency Distribution".

We're going to work with the cotinine levels of smokers. Suppose we want to construct a frequency distribution with 5 classes, class width of 100, starting with the value 0. We would compute the **lower class limits** to be 0, 100, 200, 200, 400 and the **upper class limits** to be 99, 199, 299, 399 and 499. (Keep in mind that we determined these values without the help of Excel!)

1) **Set up your worksheet:** Assuming you have followed the steps in the previous section, "Opening and Organizing Your Data", you should be ready to work with the data on Smokers in Sheet 2, which you have renamed "Smokers". Otherwise, go back and follow those steps before proceeding.

2) **Set up the Class Limit column(s):** In a blank column to the left of the data, type the name "Lower Class Limit" or "LCL" in the first cell of your selected column. In the column directly to the left of that, type in "Upper Class Limit" or "UCL". Type in the values that are listed above. (**Note:** While it is not necessary to enter the LCL into Excel, it helps us later on as we interpret our frequency table and histogram.)

	A	B	C	D
1	**SMOKER**		LCL	UCL
2	1		0	99
3	0		100	199
4	131		200	299
5	173		300	399
6	265		400	499
7	210			
8	44			
9	277			
10	32			

3) **Access Data Analysis:** Click on **Tools** and select **Data Analysis**. (If this feature is not available, go back to the instructions in Section 2-1 to learn how to add this feature to your machine.)

4) **Select Histogram:** In the Data Analysis Dialog Box, double click on **Histogram** (or select **Histogram** and click on **OK**).

Data Analysis

Analysis Tools:
Anova: Two-Factor Without Replication
Correlation
Covariance
Descriptive Statistics
Exponential Smoothing
F-Test Two-Sample for Variances
Fourier Analysis
Histogram
Moving Average
Random Number Generation

OK / Cancel / Help

5) **Work with the Histogram Dialog Box:** Follow the steps below to create the dialog box shown.

a. **Collapsing the Dialog Box:**
If you are going to fill in the ranges by selecting the appropriate cells in your worksheet, you will want to collapse the dialog box by first clicking on the "collapse" icon on the right hand side of the input boxes. ▦ This will collapse the dialog box, and allow you to move freely around your worksheet. Once you have selected your input, click on the "expand" icon ▦ to go back to the dialog box.

Histogram	✕
Input	
Input Range:	$A$2:$A$41 ▦
Bin Range:	$C$2:$C$6 ▦
☐ Labels	
Output options	
⦿ Output Range:	$E$4 ▦
○ New Worksheet Ply:	
○ New Workbook	
☐ Pareto (sorted histogram)	
☐ Cumulative Percentage	
☑ Chart Output	

Buttons: OK Cancel Help

b. **Input Range:** You need to tell Excel what data you want to sort. This is referred to as the Input Range. Since we want the frequency distribution of Cotinine Levels of Smokers, for the "Input Range," we need to tell Excel where this data can be found. You can either type the cell range in the box with a colon separating the first and last cell (for example, type **A2:A41** if your data can be found in these cells) **OR** select the cells where this data is located in your worksheet. (Refer back to Chapter 1 if you need to review how to select a range of cells.) If you are going to select the cells, you should first click on the "collapse" icon so that you can move freely around your worksheet. Once you have selected your input, click on the "expand" icon to go back to the dialog box.

c. **Bin Range:** Excel refers to the **upper class limits** as **bins.** In the **Bin Range** box, either type in the range of cells where the **upper class limit** values are located with a colon in between cell names, or select these cells in your worksheet.

d. **Output Options:** You can either have your results positioned in your current worksheet, or in a new worksheet.

i. To position your **output in the same worksheet**, click in the circle in front of **Output Range**, then click inside the entry box, and either type the cell location where you want your data to begin, or use the "collapse" icon and select an appropriate cell. Notice in the dialog box above, Excel will position the output for the frequency distribution starting in cell E4. **Note:** If you are going to overwrite data that already exists in your worksheet, a warning dialog box will appear.

ii. **New Worksheet:** If you want to have your output returned in a separate worksheet, click in front of **New Worksheet Ply**, and type in a meaningful name such as "Smoker Histogram". Make sure you click on Chart Output. Once you press **OK**, your frequency table and histogram will appear in a new worksheet which is named according to what you typed in the input box for New Worksheet Ply.

e. **Always Include Chart Output:** Since it is rare that you would want to create a frequency distribution without the connected picture, get in the habit of always clicking in front of "Chart Output". This will ensure that you create both the frequency table as well as an initial histogram.

f. **Click on OK**. You should now see a frequency distribution for your data as well as a rudimentary histogram.

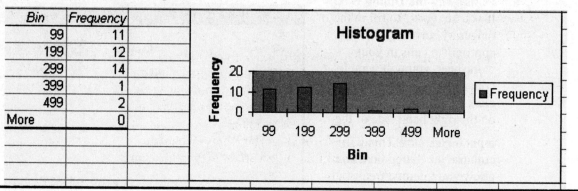

Bin	Frequency
99	11
199	12
299	14
399	1
499	2
More	0

Note:

You should remember to interpret the frequencies as the number of values that are less than or equal to the upper class limits that you typed into your bins, but greater than the previous upper class limit.

Looking at the frequency distribution for the Distribution of Cotinine Levels of Smokers, you should interpret that there are 12 ratings between 100 and 199 (less than or equal to the bin value 199, but greater than the previous bin value 99).

MODIFYING THE INITIAL FREQUENCY DISTRIBUTION

1) **Delete the row labeled "More":** Notice that when Excel created the frequency distribution, there is an extra row labeled "More." This helps ensure that you count all the data in the set, even if you mistakenly did not include a high enough upper class limit in your bins. If you included enough upper class limits, this row should show a "0". To remove this extra row, highlight only the two cells in the frequency distribution that you wish to remove. Click on Edit, highlight and click on Delete, click in the bubble by Shift Cells Up, and then click on OK. Notice that when you deleted this row, the extra column in the histogram picture also disappeared.

2) **Rename the first column of the frequency distribution:** We want the frequency distribution to clearly indicate what the data represents, so we want to create a name other than "Bin".

   a. Click on the cell where you see the name "Bin" at the top of the column, and type in the new name you would like to use. In this example, it makes more sense to type in what the values represent: "Cotinine Levels of Smokers".

   b. This name would extend over a larger region than the original column width. To rectify this, after you type in the heading, click on another cell, and then re-click on the cell containing your heading. With this cell selected, click on Format, and then click on Cells. In the dialog box that appears, click on Alignment, and click in the box in front of "Wrap Text." Then click on OK.

3) **Change the "bin" values to Class Marks:** From your textbook, you learned that a histogram should show either the class boundaries or the class marks along the horizontal axis. If you use class boundaries, they should appear under the vertical lines of the bars. If you use class marks, they should be centered under each bar. Remember that in order to have Excel create the frequency table, you needed to tell it to use the upper class limits as the 'bins". Now that you have created the initial frequency table and histogram, you need to change the bin values in the frequency table to the class marks. (In Excel, it is too complicated a procedure to use the class boundaries in the histogram!)

4) **Using Excel to compute the class marks:** If you have entered both your lower class limits and upper class limits into Excel, it is relatively easy to use a formula in Excel to automatically compute the class marks. (While you could certainly do this by hand, it's not a bad idea to get used to creating and copying formulas in Excel!)

   a. **In order to use a formula, you will reference cell addresses.** If your lower class limits started in cell C2, and your upper class limits started in cell D2, you could type the following formula in a cell below your class limits: =(c2+d2)/2. The equal sign is what tells Excel you are entering a formula. The parentheses are essential in order to perform the addition before the subtraction. When you press Enter you should see the first class mark of 49.5 in the cell where you entered the formula.

C	D
LCL	UCL
0	99
100	199
200	299
300	399
400	499
Class Marks	
=(c2+d2)/2	

   b. **Now you want to copy the formula using the fill handle.** Move the cursor to the black corner of the box containing the first class mark until it turns into a "+" sign. Now hold down the left mouse key and "pull" the handle down so that you have 4 additional boxes "selected". When you let your mouse up, you should see that your cells have been "filled" with the updated formulas which create the class marks for each of the classes using successive lower and upper class limits.

Microsoft Excel - COTININE

File Edit View Insert Format Tools Data Windo

SnagIt Window

C10   =(C2+D2)/2

	A	B	C	D	E
1	SMOKER		LCL	UCL	
2	1		0	99	
3	0		100	199	
4	131		200	299	
5	173		300	399	
6	265		400	499	
7	210				
8	44				
9	277		Class Marks		
10	32		49.5		
11	3				
12	35				
13	112				
14	477				
15	289				
16	227				
17	103				
18	222				

5) **Type or copy these class marks into your frequency table:** You can now either type your class mark values into your existing frequency table, or you can copy and paste these values into the table. (**Note: Since the values you want to use were created by a formula, if you use copy and paste, you must use "Paste Special" from the tool bar, and then paste only the values.**)

6) **Your histogram changes as you change the frequency table:** Notice that when you changed the values in your frequency table, the values also changed on your histogram. Your worksheet should now look like the following picture.

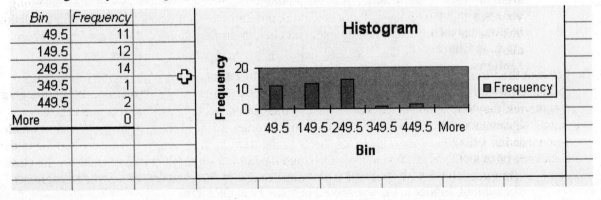

Bin	Frequency
49.5	11
149.5	12
249.5	14
349.5	1
449.5	2
More	0

MODIFYING THE INITIAL HISTOGRAM

You will need to modify the histogram some to make it look the way you want it to. In particular, you will want to close the gap between the bars, re-name the horizontal axis, and give the graph an appropriate title. In addition, you will want to resize your graph. Instructions to complete the revisions follow.

1) **Remove the gap between the bars:** A histogram has bars that are touching. Therefore we need to adjust the initial picture.

   a. To remove the gap between the bars on the Histogram, right click on one of the histogram bars. Then Click on **Format Data Series** in the shortcut menu that is displayed. (You can also double click on one of the histogram bars to go directly to the Format Data Series dialog box.)

   b. In the Format Data Series dialog box, click on the **Options** tab, and change the gap width value to 0.

   c. Click on **OK.**

2) **Changing the Graph Title and Horizontal Labels:** The graph title that comes up on your histogram by default is "Histogram," and the x axis label is "Bins." You will want to change the names to be more indicative of what the graph represents. There are two ways that you can change the information initially listed on your histogram.

   a. With the Chart Area selected, left click in this region, and then click on **Chart Options.**

   b. In the "Chart Title" and "Category (X) Axis" boxes, type in more

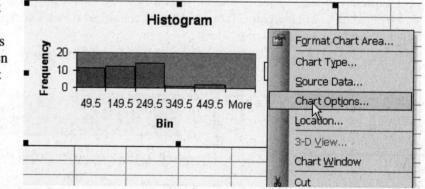

   appropriate names for the graph and the values on the horizontal axis. Your dialog screen should look similar to the one shown below. Then click **OK.**

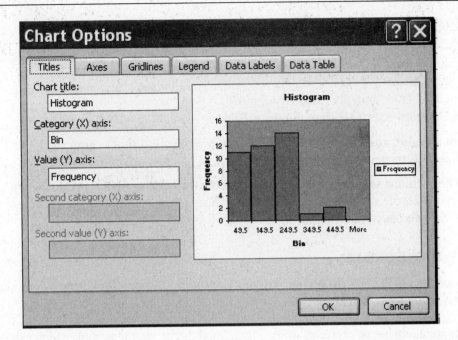

c.   **OR, alternatively:** Select the title or the axis name you want to change by clicking on it. Your title or axis name will have a "selection box" around it. To change the name, simply begin typing the new name you wish to use. Notice the text is typed up at the top of the worksheet, and the title in the box is changed once you press **Enter**.

3)   **Delete the Legend Box:** Notice you also see a Legend box entitled "Frequency." Since this adds nothing to the picture, you should delete this box. Click on the box, and press the delete key on your keyboard. **(Del)**

4)   **Resizing a Region:** You would not usually leave your picture the size it appears initially. Select the region you want to resize. Specifically, let's resize the entire "Chart Area."

a.   Notice the black "handles" on the selection box. By clicking and dragging on the corner handles, you can resize the region in both directions. By clicking and dragging on the side handles, you can stretch or shrink the region horizontally. By clicking on the middle handles at the top or bottom, you can stretch or shrink the region vertically.

b.   Move your cursor to the black box in the lower right hand corner. Your cursor should change to a double headed arrow. Left click, and holding the mouse down, drag this corner diagonally across your screen to resize the chart in both directions.

5)   **Changing the Font Size:** The last changes we will make to the histogram will be to change the font size.

a.   Move your cursor into the chart region near the horizontal axis until you see the tag "Category Axis" appear. Then right click, and click on **Format Axis**.

b.   Click on the Font tab, and make adjustments to your font size.

c.   Explore the other possibilities if you would like, and then click on **OK**.

SELECTING REGIONS IN A HISTOGRAM & ACCESSING FORMATTING OPTIONS

1)   **Position your mouse over different parts of the white box containing the histogram.** You will see "tags" come up telling you what the various regions are called. Regions include:

- Category Axis (Horizontal axis under the histogram where the "bin" values appear.)
- Chart Area
- Value Axis Title (Currently reads Frequency)
- Plot Area
- Chart Title (Initially reads Histogram)
- Category Axis Title (Initially reads Bins)

2) **Click on a region and notice that "handles" appear around that region.** This indicates that you have "selected" the region.

3) **Right click while a region is selected to access the formatting menu.** You have many options open to you in terms of what type of formatting changes you can make. We encourage you to "play" with the various options to create your own individualized picture.

TO PRACTICE THESE SKILLS

To really learn how to use Excel well, you need to practice the skills covered in the previous section several times before you "own them."

Notice that there are a number of exercises in section 2-2 and 2-3 of your text book which ask you to create a frequency distribution (section 2-2) and a histogram (section 2-3) using the data sets found in the back of your book. You should practice your skills with some of the paired sets below:

1) **Rainfall Amounts:** Exercise # 19 from section 2-2, paired with exercise # 11 from section 2-3.

2) **Nicotine in Cigarettes:** Exercise # 20 in section 2-2, paired with exercise # 12 from section 2-3.

3) **Weights of Pennies:** Exercise # 23 from section 2-2, paired with exercise # 15 from section 2-3.

SECTION 2-3: RELATIVE FREQUENCY HISTOGRAMS GIVEN A FREQUENCY DISTRIBUTION

Sometimes we already have a frequency distribution, and we want to consider the relative frequency distribution and relative frequency histogram. The steps below will take you through how you can create the relative frequency distribution by using a formula, as well as how to create the histogram using the Chart Wizard.

CREATING A RELATIVE FREQUENCY DISTRIBUTION

Suppose we had the frequency distribution from the previous section, and we wanted to create a relative frequency distribution. We need to add a column for relative frequencies to our frequency distribution. The relative frequency for a particular class will be equal to the number of values in that class divided by the total number of values in the data set. To set up this column in the frequency distribution follow the steps below.

1) **Create the Total:**

   a. In a cell under the entries in the first column of your frequency distribution, type the word "Total."

   b. Staying in the same row as where you entered "Total," click in the cell under the frequency column.

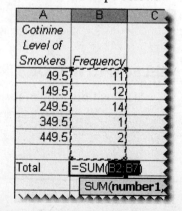

   c. Now click on the auto sum button on the toolbar. Σ ▾ You should see the following:

   d. Make sure the cells selected are the ones you wanted to add. If so, press **Enter.** If not, select the range of appropriate cells before pressing **Enter.** You should now see the total number of data values in your data set, which should be 40.

2) **Create the Relative Frequency Column:**

   a. At the top of the next column of your frequency distribution, type in the name "Relative Frequency", and press **Enter.**

   b. Re-click on the heading. Click on **Format,** and select **Cells.** Then under **Alignment,** make sure there is a check mark in front of "Wrap Text." Then click on **OK.**

   c. Since we typically want to see the relative frequencies displayed as percents, change the formatting of the column.

       i. Click on the letter at the top of the column to select the column, or select only the cells where your values will appear.

       ii. Click on **Format,** and choose **Cells.** In the **Number** tab of the dialog box, choose Percentage, and set the Decimal places to 0. Then click on **OK.**

3) **Create your formula:** In the cell immediately to the right of your data for the first class, enter the formula that will divide your frequency count for a class by the total number of data values. For example, if your first upper class limit is in cell A2, your frequency count for that class is in B2 and your total frequency count is in B8, you would type: =b2/b8, and press **Enter.** Notice that b2 (where your first frequency count can be found) is a relative address, while b8 (where the total of all frequencies can be found) is an absolute address, indicated by the $ signs. This will allow the numerator to be updated when you copy the formula, but will keep the total number of values (in this case, 40) fixed.

	A	B	C	D
1	Cotinine Level of Smokers	Frequency		
2	49.5	11	=b2/b8	
3	149.5	12		
4	249.5	14		
5	349.5	1		
6	449.5	2		
7				
8	Total	40		
9				

4) **Copy this formula:** Use the fill handle to fill in the remaining cells in your table. You should see a table similar to the following:

Cotinine Level of Smokers	Frequency	Relative Frequency
49.5	11	28%
149.5	12	30%
249.5	14	35%
349.5	1	3%
449.5	2	5%
Total	40	

CREATING A HISTOGRAM FROM A GIVEN FREQUENCY TABLE USING CHART WIZARD

Now that we have the relative frequency table, we can use the **Chart Wizard** to create the relative frequency histogram. There are a series of 4 linked dialog boxes that you need to address when using the Chart Wizard.

1) **Access the Chart Wizard:** To access the Chart Wizard, you can click on **Insert** in the menu bar and then click on **Chart**, or you can click on the **Chart Wizard** icon on the toolbar.

2) **In Step 1 of 4**, in the **Standard Types** menu, make sure that **Column** is highlighted. Then click on **Next** at the bottom of the screen. This will take you to the second dialog box.

3) **In the second dialog box (Step 2 of 4 - Chart Source Data)** enter the range where your relative frequencies are listed by selecting these cells in the worksheet. Think of range as your output values, or what you want displayed on your vertical axis. Note that the name of your sheet comes up in the front of the range, and that the cell addresses are absolute addresses.

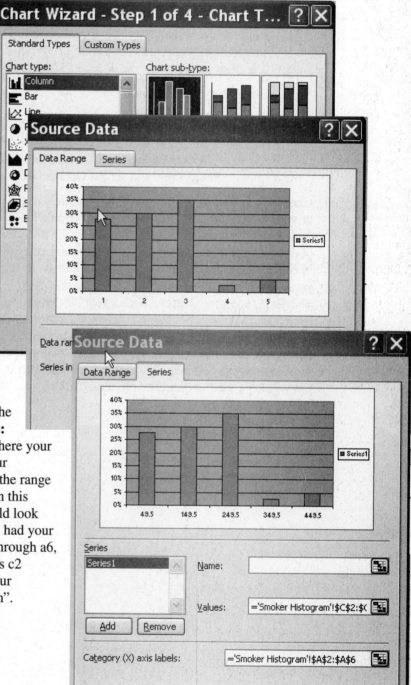

   a. In this same dialog box, click on the **Series** tab, and click in the **Category (X) axis labels** box. Enter in the range where your class marks are listed by selecting the cells in your worksheet. **Note: You MUST** select the cells where your input values are located in your worksheet. You cannot enter the range of cells separated by a colon in this region. Your dialog box should look like the one shown here if you had your class marks listed in cells a2 through a6, the relative frequencies in cells c2 through c6, and had named your worksheet "Smoker Histogram".

   b. Click on **Next**.

4) **In Step 3 of 4 - Chart Options**, type in an appropriate Title, and descriptions of your x and y axes if desired. Click on each tab at the top of the dialog box, and determine which components you wish to select. Your dialog box should look like the one shown. Click on **Next**.

5) **In Step 4 of 4 - Chart Location**, determine whether you want your chart in a new worksheet (recommended option) or inserted in an existing worksheet. Then click on **Finish.** You should now see a basic relative frequency histogram. Clearly, you will again want to modify the histogram by changing the gap width to 0, resizing the picture, etc. See the instructions for "Modifying the Initial Histogram" in section 2-2 to make appropriate modifications to your relative frequency histogram.

TO PRACTICE THESE SKILLS

To practice making relative frequency tables and relative frequency histograms, you can work with the frequency distributions you made in section 2-2 of this manual. For each of these frequency histograms, follow the directions in this section to create the relative frequency histogram, and then use the Chart Wizard to create the corresponding relative frequency histogram.

SECTION 2-4: CREATING OTHER GRAPHS

We can use Excel to create a number of other types of graphs, including Frequency Polygons, Pareto Charts and Pie Charts. The instructions contained in this section will help you create these graphs.

CREATING A FREQUENCY POLYGON

Once we have the frequency distribution for our data, we may choose to represent it graphically as a frequency polygon or a relative frequency polygon rather than a histogram. This type of graph uses line segments connected to points located directly above your class midpoint values. In the frequency polygon, the heights of the points correspond to the class frequencies. In the relative frequency polygon, the relative frequencies are represented on the vertical scale.

Let's use the data given in Table 2-8 in your book.

1) **Type the information into Excel:** We can type this information directly into Excel, using the class midpoints rather than the listed classes.

	A	B	C
1	Age	Actresses	Actors
2		0%	0%
3	25.5	37%	4%
4	35.5	40%	33%
5	45.5	16%	39%
6	55.5	3%	18%
7	65.5	3%	4%
8	75.5	3%	1%
9		0%	0%

   a. Before you enter the percentages in the columns, first select the cells, and format the cells as Percentages with 0 decimal places.

   b. Include an extra row at the beginning and end of the data so that the lines in the polygon extend down to the horizontal axis.

2) **Access the Chart Wizard:** You can click on **Insert** from the menu bar, and click on **Chart**, or click on the Chart Wizard Icon in the tool bar.

3) **Step 1 of 4:** Click on **Line** under **Chart Types**, and click on the first option in the second row of possible graph types. Then press **Enter**.

4) **Select your Data Range and Series Information:**

   a. In the **Data Range** box, enter the range where the relative frequencies are located by selecting the appropriate cells in your worksheet. **Include the column titles in this selection.**

   b. Click on the **Series** tab at the top of the dialog box, and click in the **Category (X) axis labels** box. Then select the range of cells where your class midpoints are located. **Do not include the column title here but DO include the extra rows directly above and below the class midpoints.** Then click on **Next**.

5) **Step 3 of 4 - Chart Options:** Type in an appropriate Title, and descriptions of your x and y axes if desired. Click on each tab at the top of the dialog box, and determine which components you wish to change. Then click on **Next**.

6) **Step 4 of 4 - Chart Location:** Determine whether you want your chart in a new worksheet (recommended option) or inserted in an existing worksheet. Then click on **Finish**. You will want to modify your chart by resizing it, and potentially making other changes, including changing the font sizes, to make it look more appealing.

7) **Change the Gray Background:** Right click in the Plot Area, and select **Format Plot Area.** Set both **Border** and **Area** to **None.** Then click **OK.** Your picture should look similar to the one shown below.

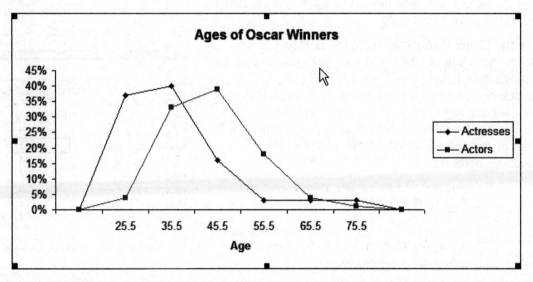

CREATING A PARETO CHART

We can use Excel to sort qualitative data in order of decreasing frequencies, and then call on the **Chart Wizard** to create a Pareto chart.

1) **Enter the following table into a new worksheet in Excel: Make sure you use adjacent columns!** You must have the data in contiguous columns in order to properly have the information sorted. Again, get in the habit of naming your worksheets. Double click on the worksheet tab, and type Pareto to indicate that this worksheet contains your work to create a Pareto chart.

Complaints Against Phone Carriers	Frequency
Rates and Services	4473
Marketing	1007
International Calling	766
Access Charges	614
Operator Services	534
Slamming	12478
Cramming	1214

2) **Sort the Data:** We first need to sort the data in order of decreasing frequency.

   a. Click on any one cell containing a value in the Frequency column. Do **NOT** select the entire column!

b. Click on **Data** in the menu bar, and then click on **Sort.** Make sure that the Sort by box shows Frequency. Click inside the circle next to **Descending.** Your dialog box should look like the following:

c. Click on **OK.** Your data should now be sorted from highest to lowest. You should see that the table labels were rearranged so that they stayed matched with the appropriate data value. Your table should look like the one shown below, where the type of complaints are listed in order of highest to lowest frequency.

3) **Use the Chart Wizard to create the graph**: Click on Insert, then click on Chart, and click on Column. Follow through the 4 linked steps in the Chart Wizard. Remember, when you enter a range, you should do so by selecting the appropriate cells in your worksheet. Notice that the name of your worksheet comes up in the front of the range, and that the cell addresses are absolute addresses.

	A	B	C
1	Complaints Against Phone Carriers	Frequency	
2	Slamming	12478	
3	Rates and Services	4473	
4	Cramming	1214	
5	Marketing	1007	
6	International Calling	766	
7	Access Charges	614	
8	Operator Services	534	
9			

a. Enter your range where your frequencies are listed as the Data range.

b. Click on the **Series** tab, and click inside the box near **Category (X) axis.** Then select the cells where your labels are located.

c. Select appropriate titles and options under the **Chart Options** menu. When you are finished, your chart should look like something like the one below. Please make sure you follow procedures to refine your original picture!

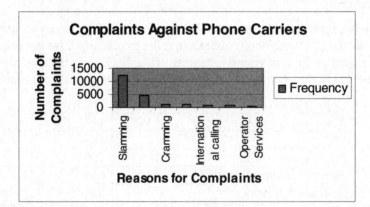

Modifying Your Pareto Chart

Once you have your basic picture, you should make further modifications to "professionalize" your chart. You should resize the chart, change the font size on the horizontal axis, and delete the "Series 1" legend box on the right hand side. You should also close the gap between the columns. Review various ways to modify the chart by re-reading material in earlier sections of this manual. A final picture of an appropriate Pareto Chart might appear as the one below.

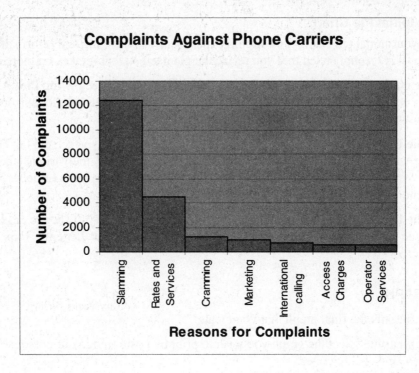

CREATING A PIE CHART

We may decide that we want to graphically show qualitative data on a pie chart, rather than as a Pareto chart. Let's work again with the information on Phone Company Complaints. The original data is listed below.

Complaints Against Phone Carriers	Frequency
Rates and Services	4473
Marketing	1007
International Calling	766
Access Charges	614
Operator Services	534
Slamming	12478
Cramming	1214

1) **Enter the Data:** Enter this information into two adjacent columns in a new worksheet, or copy it into a new worksheet if you have already entered it for the Pareto chart. **Do not leave a column in between the sources and frequencies.**

2) **Use the Chart Wizard to create the Graph:** In the **Chart Type** box (Step 1 of 4), click on **Pie**, and then select the type of pie chart you want in the **Chart sub-type** menu box. Then click on **Next** at the bottom of the screen. Follow through all the dialog boxes to create your graph. (If you haven't used the Chart Wizard yet, we recommend that you go back and follow through the detailed directions for the Pareto Chart.)

Modifying Your Initial Pie Chart

You will now see your initial pie chart. You should experiment with resizing and reformatting the various parts of your chart. It is recommended that you resize the pie itself so that it takes up more of the Chart Area.

1) **Resize the circle:** Move your cursor close to the circle until you see the tag **Plot Area** appear. Click once you see this tag appear, and you will see a rectangle with handles surround your circle. You can now resize the circle within the **Chart Area**.

2) **Format the Data Labels:** Move your cursor over the labels until you see the tag **Data Labels**. Click once you see this tag appear. You will see handles appear by the various labels. Right click, and in the menu box which appears, highlight and click on **Format Data Labels**. Experiment with the different options available.

3) **Format the Legend Box:** Move your cursor over the legend until you see the tag **Legend**. Click once you see this tag appear. Right click, and then click on **Format Legend**. Experiment with the different options.

PRINTING A GRAPH

You can elect to print only the final graph that you create.

1) **Select the Graph:** Select the graph you wish to print by positioning your cursor somewhere in the **Chart Area**, and clicking once.

2) **Preview the Graph:** Click on **File**, and in the drop down menu which appears, click on **Print Preview**. You should now see a screen shot of what your graph will look like when you print it.

   a. You can change the page margins by first clicking on **Margins,** and then positioning your cursor on the dotted page guide you want to move. Your cursor should change into a double-headed arrow bisected by a straight line segment. Hold your left click button down and drag your cursor to the position you want the margin to be. Then release the mouse. You should see your page margins change in the direction you moved.

   b. If you want to modify your graph, close the **Print Preview** window and make the changes you want to the graph.

   c. Re-check your picture in the **Print Preview** window again. When your picture is as desired, press the **Print** key found at the top of the **Print Preview** window.

TO PRACTICE THESE SKILLS

1) **Frequency Polygon:** You can work on Exercise # 12 from section 2-4.

2) **Pareto Charts:** You can work on Exercises # 13 and # 16 from section 2-4

3) **Pie Charts:** You can work on Exercises # 14 and # 15

CHAPTER 3: DESCRIBING, EXPLORING, AND COMPARING DATA 38

SECTION 3-1: OVERVIEW 38

SECTION 3-2 & 3-3 MEASURES OF CENTER AND VARIATION 38

OPENING AND ORGANIZING YOUR DATA 38

PRODUCING A SUMMARY TABLE OF STATISTICS USING DESCRIPTIVE STATISTICS 38

CREATING PARTICULAR SAMPLE STATISTICS USING THE FUNCTION WIZARD 41

COMPUTING THE MEAN FROM A FREQUENCY DISTRIBUTION 43

COMPUTING THE STANDARD DEVIATION FROM A FREQUENCY DISTRIBUTION 44

TO PRACTICE THESE SKILLS 45

SECTION 3-4: MEASURES OF RELATIVE STANDING 46

FINDING STANDARDIZED SCORES OR Z-SCORES 46

FINDING THE VALUES OF THE QUARTILES 47

FINDING THE VALUES FOR PERCENTILES 48

FINDING THE PERCENTILE FOR A PARTICULAR VALUE 49

TO PRACTICE THESE SKILLS 50

SECTION 3-5: EXPLORATORY DATA ANALYSIS 51

BOXPLOTS 51

TO PRACTICE THESE SKILLS 53

CHAPTER 3: DESCRIBING, EXPLORING, AND COMPARING DATA
SECTION 3-1: OVERVIEW

In this chapter, we will learn how to use Excel to create the basic statistics that describe important numeric characteristics of a set of data. Since technology allows us to easily create important values without having to memorize formulas or perform complex arithmetic calculations, we can spend more time making sure we know how to interpret and use these values to understand our data.

SECTION 3-2 & 3-3 MEASURES OF CENTER AND VARIATION

OPENING AND ORGANIZING YOUR DATA

Developing Good Habits: Again, you should get in the habit of preserving your original data in one of the sheets in your workbook, and then working with the specific data of interest in another worksheet. This way, you can always easily retrieve it if something goes awry as you work with specific parts of the original data. You should also clearly name the worksheet tabs so that it is clear what is contained in each sheet.

We will again work with the data from **Data Set 4 in Appendix B** (Measured Cotinine Levels in Three Groups). Follow the key steps below. The details for these steps are given in Chapter 2, section 2-2.

1) **Load the data into Excel.**

2) **Preserve your original data in a clearly marked worksheet.**

3) **Select the data you will be working with.** (We will work with the entire data set – Smoker, ETS and NOETS.)

4) **Copy the selected data to a new worksheet.**

5) **Rename this worksheet to indicate the work you are completing there.**

PRODUCING A SUMMARY TABLE OF STATISTICS USING DESCRIPTIVE STATISTICS

1) **Access the Data Analysis feature: Click on Tools in the menu bar, and then click on Data Analysis.**

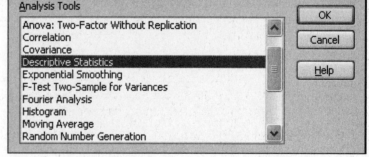

Note:

If **Data Analysis** does not show up as an option, you need to load this as an Add-In in Excel. Follow the directions in section 2-1 of this manual.

2) **Select Descriptive Statistics:** In the Data Analysis Dialog Box, double click on Descriptive Statistics (or select Descirptive Statistics and click on OK).

3) **Work with the Descriptive Statistics Dialog Box:** Suppose we want the statistics on the data for smokers.

a. **Input Range:** You need to tell Excel what data you want to use. Suppose we want to find the descriptive statistics for Cotinine Levels of Smokers. We need to either type in the beginning and ending cells where your data is located, separating the 2 cells with a colon, or select these cells in your worksheet. In the worksheet used to create this manual, this data could be found in column A, in cells A2 through A 41. Therefore, we could type A2:A41 or we could select this range of cells.

Descriptive Statistics

Input

Input Range: $A$2:$A$41

Grouped By: ⦿ Columns ◯ Rows

☐ Labels in first row

Output options

⦿ Output Range: $G$2

◯ New Worksheet Ply:

◯ New Workbook

☑ Summary statistics

☐ Confidence Level for Mean: 95 %

☐ Kth Largest: 1

☐ Kth Smallest: 1

[OK] [Cancel] [Help]

b. **Output Range:** You can either have your results positioned in your current worksheet, or in a new worksheet.

c. **To position your output in the same worksheet,** click in the circle in front of Output Range, and then click inside the entry box, and either type the cell location where you want your data to begin, or use the "collapse" icon and select an appropriate cell. Notice in the n the worksheet used to create this manual, the statistical output would begin in cell G2.

d. **New Worksheet:** If you want your ouptue returned in a separate worksheet, click in front of New Worksheet Ply, and type in a meaninful name such as "Smoker Descriptive".

e. **Choose Summary Statistics:** Make sure you choose the Summary Statistics option, and then click on **OK**. You will see your Summary Statistics displayed starting in the cell you indicated on your worksheet.

4) **Adjust your initial output:** Clearly, there are adjustments that need to be made to the initial output from Excel.

a. **Resize your columns:** You want to be able to see the full words in the first column. Move your cursor to the right hand side of the column name where you want to resize. Your cursor will turn into a double headed arrow with a straight line segment through it. When it does, double click, and your column will be

G	H	I
Column1		
Mean	172.475	
Standard E	18.89434	
Median	170	
Mode	1	
Standard D	119.4983	
Sample Va	14279.85	
Kurtosis	0.519621	
Skewness	0.587929	
Range	491	
Minimum	0	
Maximum	491	
Sum	6899	
Count	40	

G	H
Column1	
Mean	172.475
Standard Error	18.89434
Median	170
Mode	1
Standard Deviation	119.4983
Sample Variance	14279.85
Kurtosis	0.519621

automatically sized to fit the longest word in the column.

b. **Change the Title and Wrap the Text:** Click on the title Column 1 and type in "Statistics for Cotinine Levels in Smokers." Click on another cell, and then click back on your title, and format the cell to wrap the text. To do this, click on Format, choose Cells, then select Alignment, and click in the box next to Wrap Text.

c. **Delete measures that you don't need:** You may want to delete some of the standard choices from your list. Select the two cells that contain the information on Kurtosis, and press Delete. Do the same for Standard Error.

d. **Sort the data:** You now have empty rows in your table. Select all the information in the two columns of your table. Do not include the title in your selection box. Then click on the Sort icon from A to Z ![A-Z sort icon]. You will now see the information presented in alphabetical order, with the extra rows eliminated.

e. **Remove the bottom bar:** To remove the bottom bar, select the two cells containing the bottom border. Click on the down arrow by the frame icon, ![frame icon] and select the No Border icon in the displayed table ![no border icon]. Your statistics should appear as shown in the table.

Interpreting the Output in the Descriptive Output Table

Below is a brief description of each of the measures included in Descriptive Statistics.

- **Mean:** The arithmetic average of the numbers in your data set.

- **Standard Error:** This is computed by using the formula $S / \sqrt{n}$ where S is the sample standard deviation and n is the number of observations.

- **Median:** This is the data value that splits the distribution in half. To determine the value of the median, the observations are first arranged in either ascending or descending order. If the number of observations is even, the median is found by taking the arithmetic average of the two middle values. If the number of observations is odd, then the median is the middle observation.

- **Mode:** This is the observation value associated with the highest frequency. **Caution:** Three situations are possible regarding the mode: 1) if all values occur only once in a distribution, Excel will return #N/A. 2) If a variable has only one mode, Excel will return that value. 3) If a variable has more than one mode, Excel will still return only one value. The value used will be the one associated with the modal value that occurs first in the data set. To check the accuracy of the mode, it would be wise to create a frequency distribution.

- **Standard Deviation:** This is computed using the formula: $S = \sqrt{\dfrac{\sum (X - \overline{X})^2}{n-1}}$

- **Sample Variance:** This is the standard deviation squared.

- **Kurtosis:** This number describes a distribution with respect to its flatness or peaked ness as compared to a normal distribution. A negative value characterizes a relatively flat distribution. A positive value characterizes a relatively peaked distribution.

- **Skewness:** This number characterizes the asymmetry of a distribution. Negative skew indicates that the longer tail extends in the direction of low values in the distribution. Positive skew indicates that the longer tail extends in the direction of the high values.

- **Range:** The minimum value is subtracted from the maximum value.

- **Minimum:** The lowest value occurring in the data set.

- **Maximum:** The highest value occurring in the data set.

- **Sum:** The sum of the values in the data set.

- **Count:** The number of values in the data set.

CREATING PARTICULAR SAMPLE STATISTICS USING THE FUNCTION WIZARD

If you just want to know particular values, without producing the entire table of Descriptive Statistics, you can use the **Function** wizard, and select just the options that you want to use.

1) **Decide which values you want to compute**: We will compute the Mean, Median, Mode and Midrange for **Data Set 4 in Appendix B**. We will find the values for each of the groups contained in that data set: Smokers, ETS, NOETS.

2) **Copy the data on Smokers, ETS and NOETS into a new worksheet:** It is best to have this data in contiguous columns, with no blank columns in between.

E	F	G	H	I
	Smokers	**ETS**	**NOETS**	
Mean				
Median				
Mode				
Midrange				

3) **Create column and row headings:** Type in the information shown to the right.

4) **Create the first Mean for Smokers:**
   a. Select the cell next to **Mean**, and directly under the title **Smoker**. In your menu bar, click on **Insert**, and then click on **Function**. In the dialog box that appears, type the word Mean in the "Search for function:" box, and click **GO**. You should now see the work **Average** highlighted in the "Select a function" box.. Click on **OK**.

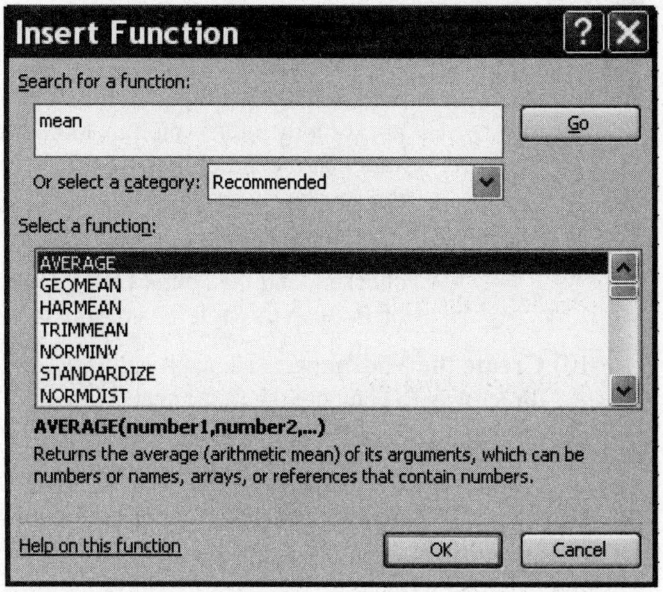

Insert Function

Search for a function:

mean   Go

Or select a category: Recommended

Select a function:

AVERAGE
GEOMEAN
HARMEAN
TRIMMEAN
NORMINV
STANDARDIZE
NORMDIST

AVERAGE(number1,number2,...)
Returns the average (arithmetic mean) of its arguments, which can be numbers or names, arrays, or references that contain numbers.

Help on this function   OK   Cancel

   b. **Fill in the Function Arguments Box:** On the line with the input box by **Number 1**, click on the icon to collapse the box, and select the data in your worksheet for which you wish to compute the mean. In the worksheet used for this manual, that data was in cells A2:A41. Then click on the icon to re-expand your dialog box. Click on **OK**. You should see the average for your data returned in the cell you had your cursor positioned in on your worksheet.

5) **Copy the Formula to other cells:** If your data for ETS and NOETS is in contiguous columns you can just copy the formula from your **Smokers Mean** cell to the cells representing **ETS Mean** and **NOETS Mean.** To do this, you can either:

    a. Use the fill handle, or

    b. You can select the box containing the mean of smokers, click on **Edit,** and select **Copy.** Then select the adjacent two cells in your worksheet (where the information for ETS Mean and NOETS Mean should appear), and then click on **Edit, Paste.** To remove the "selection" box around your initial cell, press **Esc.**

6) **If you have blank columns between the data:** In the event that your data are not contiguous, you will need to repeat the procedures from # 5, but each time select the appropriate cells for the average you are looking for.

7) **Create the Median for Smokers:** Move your cursor to the cell for **Smokers Median**. Again, click on **Insert**, and select **Function**. This time, type the word Median in the "Search for function" box and click on **GO.** You should see the function highlighted in the "Select a function" box. Click **OK.** Again, in the **Number1** box, select, or type in the range of cells where the data for Smokers is found. Then click on **OK.** If your columns are contiguous, you can merely copy the formula from this cell to the other two cells you want to fill.

8) **Create the Mode:** Move your cursor to the cell for **Smokers Mode**. Follow the same procedures as above, but type **Mode** in your "Search for function" box. Notice that after you have completed entering the cells where the data for smokers is found, and have clicked on **OK,** only one mode is returned. Unfortunately Excel will only produce one mode, even if the data is multi-modal.

9) **Consider all Modes:** To make sure you have considered all modes, you can arrange the data in order, and visually look for other modes. To do this,

    a. Select the cells where your data for Smokers is located and select **Data, Sort.** Excel will return a **Sort Warning** box, as shown. Since you do not want the other columns to be affected, make sure that the bullet in front of "Continue with the current selection" and click on **Sort.**

    b. In the **Sort** Dialog box that appears, make sure that the bullet near **Ascending** is checked, and then click **OK.** Looking at the sorted data, you see that there are 2 values of 1. Scan through the column to see if any other data appears twice.

10) **Create the Midranges:** The next values we want are the midrange values. Excel does not have a direct way to compute the midrange. However, you can easily create a formula that will return the midrange.

    a. **Sort each column:** Follow the instructions in Step 9 to sort the values in **ascending order** in each of the other columns.

    b. **Create a formula:** Since your data is in ascending order, your minimum value should be at the top of the column, and your maximum value should be at the bottom. Assuming your data on smokers is in cells A2 to A41, you would type in the formula: **=(A41+A2)/2** into the cell where you want the Midrange of Smokers to appear.

Remember, the = sign tells Excel that you are entering a formula. The parentheses are necessary in order to preserve the order of operations. Now press **Enter.**

   c. **Copy your formula to your other cells:** Again, as long as your data is in contiguous columns, you can copy the formula in your **Smokers Midrange** cell to the other two cells. Otherwise, you can repeat the procedures for the other two values.

11) **Your Table:** Your table should now look something like the one shown. You may have more or less decimal places. To set the number of decimal places shown, select the cells that contain values, then click on **Format, Cells,** and choose **Number.** Change the number of decimal places to the number you want to use, and then click on **OK.**

	Smokers	ETS	NOETS
Mean	172	61	16
Median	170	2	0
Mode	1	1	0
Midrange	245.5	275.5	154.5

12) **Other functions:** If you want to include a line for the **sample standard deviation** of your data, you would access the function **STDEV**, and again, select the cells containing the data for which you wanted to compute the sample standard deviation. Likewise, if you wanted the **sample variance**, you would access the function **VAR**, and select the cells containing the data for which you wanted to compute the sample variance. **NOTE:** Excel also allows you to compute the standard deviation and variance of a population, by selecting **STDEVP** and **VARP.**

COMPUTING THE MEAN FROM A FREQUENCY DISTRIBUTION

We can use Excel to create the mean from a frequency distribution, such as the one in Table 3-2 in your book. We will use the information we already have available on Cotinine Levels of Smokers. If you saved your work for your Smoker Histogram, you can begin with that. Otherwise, type in the data as shown:

	A	B	C
1	Cotinine Level of Smokers	Frequency	
2	49.5	11	
3	149.5	12	
4	249.5	14	
5	349.5	1	
6	449.5	2	
7			
8			

You can center the information in all columns by selecting the columns and then accessing your **Center** icon in your toolbar.

1) **Create the product of class midpoint and frequency:** In the column where you will show the frequency times the class midpoint, (the 4th column in Table 3-2) you will need to enter a formula. In the row where your first frequency and class midpoint are listed, type in the formula: = (cell reference where frequency is listed) * (cell reference where class midpoint is listed). In the worksheet created for this problem, we typed in the following formula: =A2*B2. Then press **Enter.**

2) **Copy the formula down the product column:** Position your cursor back in the cell where the product of the frequency and class midpoint appears. Then use the fill handle to copy the formula down the rest of the column.

3) **Create the necessary sums:** For the mean, you want to divide the sum of the products of the frequency and the class midpoints by the sum of the frequencies. You will need to create the sums that you want to use by following the steps below.

   a. Position your cursor in the cell directly under the last frequency, and select the summation key Σ ▾ in your toolbar. You want to add the values in the frequency column. This column should be automatically selected when you press the summation key, but if it is not, select the range of cells you want to sum up. Then press **Enter.**

b. Position your cursor in the cell directly under your last product of (frequency * class midpoint), and select the summation key from your toolbar $\Sigma \cdot$. You want to add the values of the products in this column. Again, the column should be automatically selected when you press the summation key, but if it is not, select the range of cells you want. Then press **Enter.**

4) **Create a formula for the mean:** To find the mean, you want to divide the sum from your column of products by the sum of the frequencies. In the worksheet we created, we would represent this by the formula: = C7/B7. Notice that we entered this formula in cell g7, to create a result of 177. Your worksheet should look similar to the following:

	A	B	C	D	E
1	Class Midpoint for Cotinine Level of Smokers	Frequency	Frequency times Class Midpoint		
2	49.5	11	544.5		
3	149.5	12	1794		
4	249.5	14	3493		
5	349.5	1	349.5		
6	449.5	2	899		
7	Total	40	7080		
8				x bar	177
9					
10					

5) **Changing the number of decimal places:** If you want to include more decimal places, you can select all the cells where you want more decimal places to appear, and select **Format,** then select **Cells, Number,** and set the number of decimal places to the number you wish to use. Then press **OK**.

COMPUTING THE STANDARD DEVIATION FROM A FREQUENCY DISTRIBUTION

Let's use the frequency distribution given in your book for Cotinine levels. As with finding the mean, you need to create a column that represents the frequency * the class midpoint. You also need to create a column for the frequency * square of the class midpoint.

1) **Create a column for Frequency times the square of the class midpoint:** Use the worksheet you set up to compute the mean of the frequency distribution if you have it saved, or create the worksheet from above, under the section "Computing the Mean from a Frequency Distribution." In the column directly after the column where you show the frequency * class midpoint, create another column using a formula that shows the frequency * square of the class midpoint. For the worksheet we used, our formula for the first row of data was: =B2 * A2^2 **NOTE:** To square a value, you use the carot (^) to indicate that the next value typed is the exponent.

2) **Compute the sum of these values:** As you did to compute the mean, use the summation key to sum up the values in that column.

3) **Use the formula for Standard Deviation:** Now we want to use the formula:

$$s = \sqrt{\frac{n[\sum (f \cdot x^2)] - [\sum (f \cdot x)]^2}{n(n-1)}}$$. Think about what these values represent.

a. **The value n** is the number of values included in the computation, so for this case, it will be the sum of the frequencies.

b. We already have computed in the table the **sum of the frequencies times the squares of the class midpoints** $\sum (f \cdot x^2)$.

c. We also have the value of the **sum of the frequencies times the class midpoints** $\sum (f \cdot x)$.

d. **Creating the formula in reference to cell addresses:** We need to create a formula that references the cells where these values are found. For the worksheet shown below, the formula we would enter to compute the value we want to find is: =sqrt(((b7*d7)-c7^2)/(b7*(b7-1))). Make sure you think about where you need to put parentheses in order to maintain the appropriate order of operations. You need to make sure that you take the square root of the entire expression. Then you want to make sure you group the terms that represent the numerator. Finally, you need to group the factors that comprise the denominator. Look carefully how the parentheses shown group the various components together. This is a difficult expression to enter into Excel, because the grouping is essential in order to arrive at the appropriate answer. Your worksheet should look like the one shown:

	A	B	C	D	E	F
1	Class Midpoint for Cotinine Level of Smokers	Frequency	Frequency times Class Midpoint	Frequency times square of class midpoint		
2	49.5	11	544.5	26952.75		
3	149.5	12	1794	268203		
4	249.5	14	3493	871503.5		
5	349.5	1	349.5	122150.25		
6	449.5	2	899	404100.5		
7	Total	40	7080	1692910		
8				x bar	177	
9						
10				Standard Dev.	106.1868	
11						
12						

TO PRACTICE THESE SKILLS

You can apply the skills learned in this section by working on the "Skills and Concepts" exercises found after section 3-2 and 3-3 in your textbook. **Notice that the data for the problems in Section 3-2 mirrors the data that is used again in the problems in Section 3-3.**

1) **To practice finding Measures of Center and Measures of Variation from data sets:** Work on the exercises 5 through 24 in both Sections 3-2 and 3-3 Basic Skills and Concepts in your textbook.

   - For exercises 5 through 20, you will need to type the data into Excel.

   - For exercises 21 through 24, you can load the data from the CD that comes with your book. As always, save your work with a file name that is indicative of the problem that you were working on.

2) **To practice finding the mean and standard deviation from a frequency distribution:** Work on exercises 25 through 28 in Sections 3-2 and 3-3 Basic Skills and Concepts in your textbook.

SECTION 3-4: MEASURES OF RELATIVE STANDING

FINDING STANDARDIZED SCORES OR Z-SCORES

When looking at a set of data, it is often useful to know how far a particular score falls from its mean. We can measure the position of a particular value with respect to the mean using z-scores. We know that if a value is more than 2 standard deviations away from the mean of the data set, it can be considered "unusual." Remember that whenever a value is below the mean, the corresponding z-score will be negative.
We will again work with the data from **Data Set 4 in Appendix B** (Measured Cotinine Levels in Three Groups). We will work specifically with the data on Smokers.

1) **Load your data into a new worksheet:** You can review detailed steps for this in Chapter 2, section 2-2. Again, if you are starting a new file, get in the habit of renaming the sheet that your original data is in "Original Data" by double clicking on the tab at the bottom of the worksheet, and typing in the name. Then copy the data for **Smokers** to a new sheet to work with, so that you always have a version of your original data to refer back to if necessary.

2) **Create a worksheet that shows the data for smokers in column A.** In order to create the z scores, we will need to know the mean and standard deviation for this data.

3) **Compute the mean and standard deviation:** In a column not contiguous to column A, use the function wizard to find the mean and standard deviation of this sample data. (You found these values back in sections 3-2 and 3-3, so refer back to those instructions if necessary.) In our worksheet, we created these values in cells D1 and D2.

4) **Create the column for Standard Score:** If your original data is in column A, position your cursor in cell B1, and type "Standard Score."

5) **Use the Standardized Function;** Position your cursor in the cell directly beneath this heading. In our worksheet that is cell B2. From the **Function** menu (accessed either through **Insert, Function**, or by pressing the function icon), type Standard score in the "Search for a function" box. Then click on **GO**. In the "Select a Function" box, you want to click on **Standardize** and then click on **OK**.

6) **Fill in the dialog box:**

   a. **For X:** In the dialog box, type in **A2** or the address where your first value is found.

   b. **For the Mean:** In the box for the **Mean**, type in the **absolute address** for the cell where your mean is found. In the worksheet we created, our mean was in cell E1, so we type in: $E$1.

   Notice that an absolute address, which will remain constant for all computations, whereas the x values will be updated when you use the fill command.

c. **For the Standard Deviation:** In the box for the **Standard_dev**, type in the **absolute address** for the cell where your standard deviation is found. In the worksheet we created, our standard deviation is in cell E2, so we type in: $E$2. Your dialog box should look like the one shown.

d. Click on **OK**. You will now see the Standardized score for the first number in your data list in cell B2.

7) **Format the Standard Score Column:** Since z-scores are normally reported to only two decimal places, you should format this column to show only two decimal places.

 a. **Click on the letter at the top of your column** for z-scores to select this whole column.

 b. Then click on **Format, Cells,** and **Number**, and then indicate 2 in the box by Decimal Places. Click on **OK**. You should now see the value for the z scores rounded to 2 decimal places. In our example, the first value in the table is 1, and we see a z score of -1.43. This indicates that the value of 1 is 1.43 standard deviations below the mean. Make sure you agree with this. The mean for our example is 172.475. The standard deviation is 119.498. If we subtract the mean from our value (1), and then divide by the standard deviation, we find that we do get a number which rounds to -1.43.

8) **Copy the formula down the column:** Use the fill handle to copy the formula down for the rest of the values in your data set. Notice that your values have z-scores ranging from –1.44 (indicating that 0 is 1.44 standard deviations below the mean) to 2.67 (indicating that 491 is 2.67 standard deviations above the mean).

	A	B	C	D	E	F
1	SMOKER	Standard Score		Mean	172	
2	1	-1.43		Standard Dev	119.4983	
3	0	-1.44				
4	131	-0.35				
5	173	0.00				
6	265	0.77				
7	210	0.31				
8	44	-1.08				
9	277	0.87				
10	32	-1.18				
11	3	-1.42				

FINDING THE VALUES OF THE QUARTILES

1) **Copy the data on Cotinine Levels for Smokers: Using the data on Cotinine Levels for Smokers, copy the data to column A of a new worksheet.**

2) **Set up a column for Quartiles:** Suppose we want to find the first, second and third quartiles for the data. In cell C1, type in "Quartile." In cells C2 through C4, type in 1, 2, and 3.

3) **Use the Quartile Function:** Move your cursor to cell D2 and click on the Function icon on your menu bar, or click on Insert, Function. In the "Search for a function" box, type Quartile. Then click on the name Quartile in the "Select a function" box, and click on **OK**.

4) **Fill in the dialog box:** In the dialog box, you need to fill in information on the Array and the Quartile.

 a. **Make sure you use absolute cell addresses in the Array box:** In the Array box, you must

Function Arguments

QUARTILE
Array a2:a41 = {1;0;131;173;265;2
Quart c2 = 1

= 86.75

Returns the quartile of a data set.

Quart is a number: minimum value = 0; 1st quartile = 1; median value = 2; 3rd quartile = 3; maximum value = 4.

Formula result = 86.75

Help on this function   OK   Cancel

enter your range of cells using absolute addresses. If you just select the cells, or just type in the range with a colon, the proper set of values will not be used when you copy the formula for the 2nd and 3rd quartiles! For our worksheet, the data appears in cells A2 through A41.

   b. In the **Quart** box, type in the cell where your first quartile appears. In our worksheet, this value of 1 appears in cell C2. Then click on **OK**. You should see that the first quartile is the value 86.75. Roughly speaking, this means that if you sorted the original data values, about 25% of the sorted values in your table would be less than or equal to 86.75. We can say that at least 25% of the sorted values will be less than or equal to 86.75 and at least 75% of the sorted values will be greater than or equal to 86.75.

5) **Use the fill handle to copy the formula:** Copy the formula down into the next two cells. You should see the values below:

6) **See that these values make sense:** To see that these values make sense in terms of your data, select column A by positioning your cursor on the A at the top of the column, and clicking once. Your entire column should now be selected. Then click on the Sort icon which shows from A to Z in your menu bar. This means that your data will be sorted from smallest to largest value.

	A	B	C	D	E
1	SMOKER		Quartiles		
2	1		1	86.75	
3	0		2	170	
4	131		3	250.75	
5	173				
6	265				

Note:

Notice that there are 10 values which are less than or equal to 86.75. Since 10/40 represents 25%, we can see that **at least** 25% of the data values are less than or equal to 86.75. If we counted the values that were less than or equal to 170, we would find that 20 out of the 40 values fall into this category. Doing the division, we find that 20/40 = 50 %. We can say that **at least** 50% of the values are less than or equal to 170.

There are 30 values which are less than 250.75. Since 30/40 is 75%, we can see that **at least** 75% of the data values are less than or equal to 250.75.

FINDING THE VALUES FOR PERCENTILES

Let's now calculate the 10th, 20th, 30th,....,90th percentiles. These values will correspond to the 1st, 2nd, 3rd,...., 9th deciles.

1) **Create a column for Percentile:** Using our previous worksheet where we computed the quartiles, position your cursor in cell F1 and type in "Percentile."

2) **Enter the 9 Percentiles as decimals:** Starting in cell F2, enter the values 0.1, 0.2, 0.3, 0.4,....., 0.9. (Notice that Excel requires that you show the decimal form of the percentile that you want to find. 0.1 refers to the 10th percentile, since 0.1 is the decimal form of 10%.)

3) **Access the Percentile Function:** Move to cell G2, and click on the Function icon on the main toolbar. Type in Percentile in the "Search for a function" box, then click on Percentile from the "Select a function" box and click on **OK**.

4) **Fill in the Dialog Box:** In the dialog box, you need to fill in information on Array and K:

a. **Use absolute cell addresses for Array:** Indicate that the **Array** is stored in cells A2 to A41, again **using absolute addresses!** If you do not use the absolute addresses, when you copy your formula, you array values will change.

b. Indicate that the k value (for the kth percentile) is stored in cell F2. Click on **OK.**

5) **Copy the formula:** You should now see the value 15.6 in cell G2. Use the fill handle to copy the formula down the rest of the column. You should see the following table:

6) **Interpreting the values:** Remember that the values can be used to talk about what percent of values in the data set are less than or equal to these values. For example, at least 60% of the values in the table should be less than or equal to 208.8. If you order your data for Cotinine level of Smokers, you can see that there are 24 values less than 208.8. If you divided 24/40, you would see that there are 60% of the values in the table which are less than or equal to 208.8.

F	G	H
Percentiles		
0.1	15.6	
0.2	47.2	
0.3	109.3	
0.4	130.6	
0.5	170	
0.6	208.8	
0.7	237.3	
0.8	265.2	
0.9	289.1	

7) **Other percentile values:** If you had a different percentile, you would just enter the appropriate decimal value for k. For example, if you wanted the 35th percentile, you would use a k value of .35.

FINDING THE PERCENTILE FOR A PARTICULAR VALUE

Sometimes you want to know the percentile that corresponds to a particular value in a data set. You can use features of Excel to help you quickly determine this, particularly if your data set is large. For a small data set, it is probably just as easy to follow the procedures outlined in your book without calling on Excel. **Let's consider finding the percentile for the cotinine level of 112 in Smokers.**

1) **Sort the Data:** The first thing you want to do is sort the data you are working with. If we are using the cotinine levels of smokers, we would want to sort the data in that column. Copy the data for cotinine levels of smokers into a new worksheet. Then click on the letter at the top of your column to select the entire column. Then press the **Sort Ascending** key in your toolbar [A↓Z], or select **Data, Sort,** and then make sure that the bullet in front of Ascending is checked. Your data should then be sorted from lowest to highest value.

2) **Determine how many values you have less than your designated value:** Position your cursor in a cell in a nearby column, and press **Insert, Function,** or click on the function icon in your toolbar. You want to count the number of values less than 112 in this case, so select **Count,** and click **OK.**

3) **Fill in the dialog box:** In the **Value1** box, enter the range of cells which contain numbers less than 112 in your sorted column, or select those cells with your cursor. Then click on **OK.** You should see the value 12 returned in your cell. This tells you that there are 12 values less than 112 in your sorted list. **It is imperative that you work with a sorted list!**

4) **Compute your percentile:** Since there are 40 values in the data set, you want to divide the number of values less than 112 by 40, and then multiply the result by 100. You can set up a formula in Excel to accomplish this. You would enter the formula as follows: = (cell where count is found)/40*100. In the worksheet we used, we used the formula: =C2/40*100, and received the answer of 30. This means that 112 is the 30th percentile.

5) **Using more values:** If you were going to compute the percentile for a number of values, you could set up a worksheet as shown below. Though you have to change the cells within the **Count** function

each time, since the number of values you want to count changes with each value for which you are finding the percentile, once you set up the formula for your last column in the first row, you can copy the formula to the other rows. The table below shows the percentiles for the values 112, 210 and 290.

	A	B	C	D	E
1	SMOKER	Value	# of Values < Given Value	Percentile	
2	0	112	12	30	
3	1	210	24	60	
4	1	290	36	90	
5	3				
6	17				

TO PRACTICE THESE SKILLS

1) **To practice finding z-scores:** You can work on the exercise 9 from Section 3-4 Basic Skills and Concepts in your textbook.

2) **To practice finding percentiles and quartiles:** You can work on with the other data contained in the Cotinine Data set.

SECTION 3-5: EXPLORATORY DATA ANALYSIS

In order to work on the material in this section, you will need to load the **DDXL** Add-In that is supplied with your book. Instructions for loading DDXL can be found in section 2-1 of this manual.

BOXPLOTS

Excel is not designed to generate boxplots. You can use the **DDXL** Add-In that is supplied with your book to generate this type of graph.

1) **Open a worksheet where you have the data on Cotinine Levels for Smokers entered in column A.**

2) **Make note of the range where your actual values are stored:** If you entered the name Cotinine levels in cell A1, and the data directly below this, your values will be contained in cells A2:A41.

3) **Click on the DDXL command:** On the main menu bar, you should see DDXL.

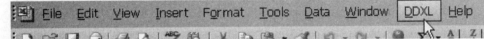

4) **Select Charts and Plots:** Once you have clicked on DDXL, you will see a number of options. Click on the Charts and Plots option. This will take you to a Charts and Plots Dialog Box.

5) **Select Boxplots:** In the Charts and Plots Dialog Box, select Boxplot.

6) **Click on the pencil icon:** At the bottom of the dialog box, you will see a pencil icon. Click on this icon. You will now see your cursor in the box immediately to the left of the icon.

7) **Enter the cells where your data is stored:** Type in the cells where the data you want to use is stored in your spreadsheet. In this case, you should type in A2:A41. Then click on **OK** at the bottom right hand side of the screen. You will then be taken to the DDXL screen, which should appear as the one displayed on the next page.

8) **Familiarize yourself with the output:** Notice that the box plot is in the upper left hand corner, and that the Summary Statistics box can be found directly below that. Click on the triangle in the upper left corner of the Box plot screen, and select Plot Scale. Notice that you can change the settings for the Y axis as shown in the Scale Plot window. Make changes if you so desire, and then click on **OK.**

9) **Create a larger picture:** Click and hold on the diamond shape in the lower right hand corner of the Box plot screen ◈ and drag this corner out to create a larger graph if you would like.

10) **Enlarge the Summary Statistics Window:** Click anywhere in the Summary Statistics window to activate that screen. Again, click and hold on the diamond ◈ in the lower right hand corner of this box, and drag to the right to expand the amount of the box that can be seen. You want to be able to see all the scores shown in the chart on the next page. Notice that different programs can produce slightly different values for the different percentiles. We found the 3rd quartile (or 75th percentile) in Excel to be 250.75, while DDXL shows it to be 251.5. Though you may see some inconsistency in the exact values returned between programs, any value you get should be in the same general ballpark!

▷ Summary Statistics (Scroll Right)										
Count	**Mean**	**Median**	**St.Dev.**	**Variance**	**Range**	**Min**	**Max**	**IQR**	**25th%**	**75th%**
40	172.475	170	119.498	14279.846	491	0	491	165	86.5	251.5

TO PRACTICE THESE SKILLS

You can apply the technology skills covered in this section by working through the exercises 7 through 12 from Section 3-5 Basic Skills and Concepts of your textbook. Remember that you may have already loaded some data from the CD into an Excel workbook for work from a previous section. You can open this file, and create a new worksheet within the file for any additional work you do with this particular data set.

CHAPTER 4: PROBABILITY

SECTION 4-1: OVERVIEW 55

SECTION 4-2: PIVOT TABLES 55

CREATING A PIVOT TABLE: 55

PROBABILITIES 58

TO PRACTICE THESE SKILLS 58

SECTION 4-3: GENERATING RANDOM NUMBERS 59

SECTION 4-4: PROBABILITIES THROUGH SIMULATION 60

CREATING A SIMULATION 60

TO PRACTICE THESE SKILLS 61

SECTION 4-5: FACTORIAL 61

SECTION 4-6: PERMUTATIONS AND COMBINATIONS: 52

TO PRACTICE THESE SKILLS 63

SECTION 4-1: OVERVIEW

This chapter in your textbook covers the basic definitions and concepts of probability. While many of those concepts are straightforward and can be done without the use of technology there are some features of Excel that can be utilized while working through the material found in this chapter. The following list contains an overview of the topics and functions that will be introduced within this chapter.

PIVOT TABLE
This feature is used to generate a worksheet table that summarizes data from a data list. This allows you to obtain category counts.

RANDBETWEEN
This function creates columns of random numbers that fall between the numbers you specify. A new random number is returned every time the worksheet is calculated. The function appears in the following format: **RANDBETWEEN (bottom, top)** where bottom is the smallest integer RANDBETWEEN will return and top is the largest integer RANDBETWEEN will return.

FACT
This function returns the factorial of a number. The factorial of a number is equal to $1 \cdot 2 \cdot 3 \cdot \ldots \cdot$ number. The function appears in the following format:
FACT(number) where number refers to the nonnegative number you want the factorial of. If the number is not an integer, it is truncated.

PERMUT
This function returns the number of permutations for a given number of objects that can be selected from a larger group of objects. A permutation is any set or subset of objects or events where internal order is significant. The function appears in the following format:
PERMUT(number, number_chosen) where number is an integer that describes the number of objects and number_chosen is an integer that describes the number of objects in each permutation.

COMBIN
This function returns the number of combinations for a given number of items. Use COMBIN to determine the total possible number of groups for a given number of items. The function appears in the following format:
COMBIN (number, number_chosen) where number is the number of items and number_chosen is the number of items in each combination.

SECTION 4-2: PIVOT TABLES

In Excel it is possible to generate a table, called a **pivot table,** which can be used to summarize both quantitative and qualitative variables contained within a database. A **pivot table** provides us with a mechanism for creating subgroups (or samples), and gives us the ability to find sums, counts, averages, standard deviation and variance of both a sample and population.

CREATING A PIVOT TABLE:

We will create a pivot table using the data from Data Set 16 in Appendix B. This data can also be found on the CD-ROM data disk that accompanies your textbook. Begin by opening Excel. Open the data file "HOME SALES".

Excel provides a **PivotTable Wizard** similar to the ChartWizard you have used to generate tables and graphs.

To create a Pivot Table:

1) On the menu bar click on **Data**, highlight **Pivot Table and Pivot Chart Report** and click. The Pivot Table and PivotChart Wizard – **Step 1 of 3 dialog box** will appear.

   a. Make sure that both **Microsoft Excel list or database** and **PivotTable** are selected.

   b. Click on **Next**.

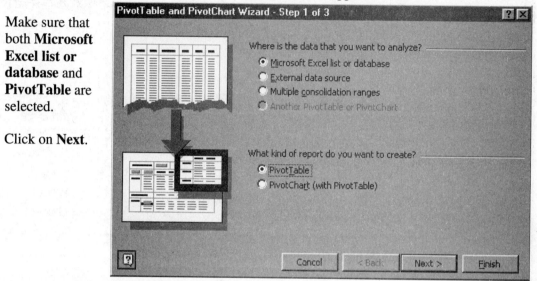

2) In the **Step 2 of 3 dialog box** indicate the range of cells that contains the data you want to use. This may already be indicated when the dialog box opens.

   a. To select your data drag the cursor from cell A1 to cell J63. In the **Range window** you should see $A$1:$J$63.

   b. Click on **Next**.

3) The **third dialog box** of the PivotTable and PivotChart Wizard will ask if you want the Pivot Table in a new worksheet or in the existing worksheet. Once you have made your choice click on **Finish**.

4) You will see a screen similar to the one shown on the following page.

   a. The small Pivot Table box displays a graphical layout of an empty pivot table along with the field buttons that correspond to the columns in your Excel worksheet.

   b. **You can specify** the layout of the pivot table by **dragging fields from the PivotTable Field** list into any of the four areas shown.

c.   You can drag as may field buttons as you like. If you make a mistake or wish to change the field button you moved, simply drag the field button back to the Pivot Table box. You can rearrange the Pivot Table as necessary.

To create a Pivot Table (continued):

5)   To help you create your first Pivot Table

a.   Drag LIVING AREA to **Drop Page Fields Here**

b.   Drag ROOMS to **Drop Column Fields Here**

c.   Drag BEDRMS to **Drop Row Fields Here**

d.   Drag LIVINGAREA to **Drop Data Fields Here**

	A	B	C	D	E	F	G	H	I	J
1	LIVINGAREA	(All)								
2										
3	Sum of LIVINGAREA	ROOMS								
4	BEDRMS	5	6	7	8	9	10	11	12	Grand Total
5	3	10	36	143	37	20	24			270
6	4			57	376	113		64		610
7	5					43	58		39	140
8	6						36			36
9	Grand Total	10	36	200	413	176	118	64	39	1056

The table you created shows the number of total rooms and number of bedrooms broken down by the total living area of the houses sold. For example the total living space of all houses with 5 rooms and 3 bedrooms is 1000 sq ft. The total living space for all houses sold with 6 rooms and 3 bedrooms is 3600 sq. ft. Note that we do not know how many houses were sold in this category, nor the square footage of each house, just the total square footage of all houses that fit that description. A few other examples of information contained in the pivot table include the fact that none of the houses with 7 rooms had 5 or 6 bedrooms. Houses with 12 rooms in my data set all had 5 bedrooms.

Note:

Once you have created a Pivot table you can refine it. Right click on any cell within the Pivot Table for a list of options. The Wizard option will take you back to the Pivot Table Wizard used to create your existing table. Double click on the **field button** within the Pivot Table to open the Pivot Table Field dialog box for options concerning fields. As always, we encourage you to explore and experiment with the different options available to you.

PROBABILITIES

Using a pivot table allows you to determine probability information more easily than sorting through the information presented to you in the original Excel worksheet.

TO PRACTICE THESE SKILLS

Use the information presented in the following Excel worksheet to create a Pivot Table that will display the total number of credits broken down by Religion and Major for
   a) male and female students combined
   b) only female students
   c) only male students

	A	B	C	D	E	F
1	SEX	AGE	MAJOR	CREDITS	GPA	RELIGION
2	M	22	Liberal Arts	19	2.5	Jewish
3	M	23	Computer Science	14	3.7	Protestant
4	F	19	Criminal Justice	17	3.8	Protestant
5	M	22	Mathematics	18	2.4	Protestant
6	F	21	English	13	2.5	Jewish
7	F	23	Liberal Arts	18	3	Catholic
8	F	22	Liberal Arts	17	3.2	Catholic
9	M	22	Liberal Arts	18	3.6	Protestant
10	M	22	Education	13	3.5	Catholic
11	F	21	English	17	2.7	Protestant
12	F	22	Criminal Justice	15	2.5	Catholic
13	M	23	Computer Science	15	3.9	Jewish
14	F	22	Computer Science	17	3.3	Jewish
15	M	21	Engineering	12	2.5	Protestant
16	F	22	Engineering	18	4	Protestant
17	F	21	Mathematics	16	3.6	Catholic
18	F	22	Liberal Arts	18	3.4	Jewish
19	M	19	English	17	2.9	Jewish
20	M	21	Criminal Justice	17	3	Catholic
21	F	20	Engineering	16	2.8	Catholic

SECTION 4-3: GENERATING RANDOM NUMBERS

Begin by making cell A1 your active cell.

1) To generate a random set of numbers in Excel, click on **Insert**, highlight **Function** and click. The following **Insert Function Dialog box** will open. We have already made use of Excel's built in Statistical functions in previous sections.

    a. **Click on the down arrow** ▼ by "select a category". Highlight **Math & Trig** as seen on the right.

    b. Scroll through the list of function names and highlight **RANDBETWEEN.**

    c. Click **OK**.

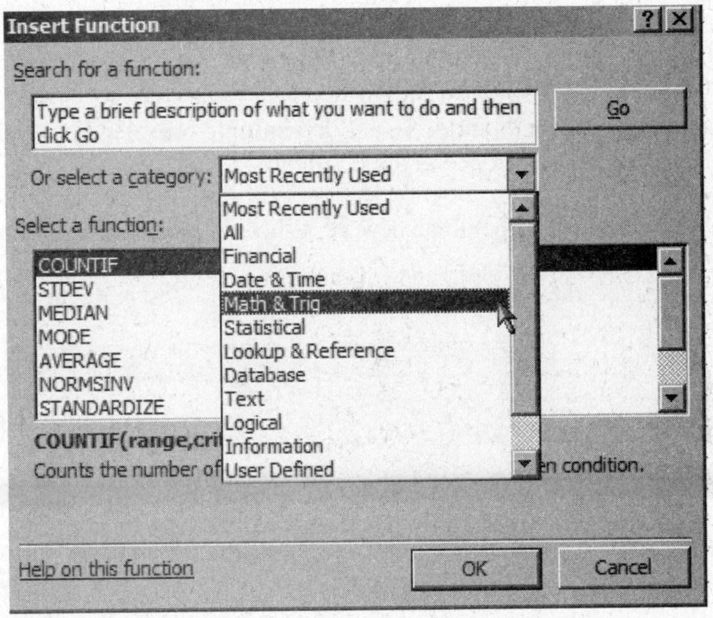

2) The dialog box (see below) opens in your Excel worksheet. Fill in the lowest and highest values between which the random number will fall. For example, if you wish to generate a list of possible values that can occur when you roll a single die then the bottom number would be 1 and the top value would be 6. Click **OK**.

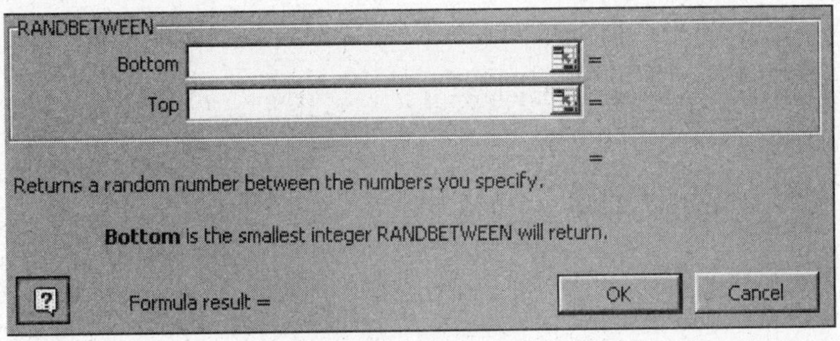

A random number between 1 and 6 should now appear in cell A1. To generate 25 such numbers copy (by dragging) this cell entry to cell A25. When you have completed this task you should have 25 entries in the first column, each between 1 and 6. A new random number is returned every time the worksheet is calculated.

SECTION 4-4: PROBABILITIES THROUGH SIMULATION

The goal of every statistical study is to collect data and to use that data to make a decision. Often collecting data or repeating a trial a large number of times can be impractical. With the use of technology we can often simulate an event.

Your textbook defines a **simulation** as follows: *"A simulation of a procedure is a process that behaves in the same way as the procedure, so that similar results are produced."*

The random number generator can be used to simulate a variety of statistical problems.

CREATING A SIMULATION

Let's consider the **Gender Selection example** presented in section 4-6 of your textbook. We can simulate the 100 births mentioned in the Example.

1) Begin by opening a new worksheet in Excel. Make cell A1 your active cell.

2) Use the method outlined in Section 4 – 3 to simulate 100 births. Let 0 = male and 1 = female.

3) To generate the 100 random values you can copy the information through to cell A100 or you can copy it through to cell A25. You can then copy the information into columns B, C and D. The advantage of this second method is that you can see the 100 entries in your simulation. You may notice that the entry in your first cell change. Don't worry about this. The cells will continue to change until we do our statistical analysis.

4) To determine the probability that the newborn if female, count the number of 1's in your data and divide by the total number of entries.

It might be interesting to repeat the simulation several times and compare your results.

Simulating Multiple Events

Consider a problem that requires more than one event to occur such as the probability of a specific sum when two dice are tossed as mentioned in the **Simulating Dice example** found in Section 4-6.

Suppose that for the purpose of this simulation we wish to find the P (sum of 5) when two dice are tossed 50 times. One possibility would be to toss a pair of dice 50 times and record the sums after each roll. Using technology to simulate this experiment we will produce similar results.

1) Begin a new worksheet with cell A1 as the active cell. Type "First Die" in cell A1, type "Second Die" in cell B1 and type "Sum" in cell C1.

2) **In cell A2** use the method outlined in Section 4 – 3 to generate a *column* of 50 random values between 1 and 6.

3) **In cell B2** generate a *column* of 50 random values between 1 and 6. As before, you may notice that the cell entries in your first column change. This is normal. The cells will continue to change until we do our statistical analysis.

4) **In cell C2** type **=Sum (A2 + B2)** and press **Enter**.

5) **Copy this formula** down through to cell C51. This should give you the sums of the toss of two dice. Once again you will notice that the entries in your first two columns have changed. This is normal. You should be able to see rather easily that the sum of the first two column entries is reflected in the third column.

Now it is possible to determine how many of our simulated tosses of the dice yield a sum of 5, that is, we can find P(sum of 5).

TO PRACTICE THESE SKILLS

You can practice the Excel skills learned in Sections 4-3 and 4-4 of this manual by working through the following problem.

1) Some role-playing games use dice that contain more sides than the traditional six sided dice most of us are familiar with. Assume we are playing such a game and that we are using a pair of ten sided dice which contain the numbers one through ten each die.
   a) Use the Random Number Generator to simulate rolling a pair of ten sided dice fifty times.
   b) Use the results to determine a list of possible sums from the fifty rolls of the dice.
   c) Find the probability of rolling a sum of 20.

2) Develop a simulation using Excel for Exercises 9 and 11 in the Basic Skills and Concepts for Section 4-6.

SECTION 4-5: FACTORIAL

In looking at the solution to the **Cotinine in Smokers** example found in section 4-7 of the textbook we see that we are required to multiply $3 \cdot 2 \cdot 1$. This product can be represented by 3! which is read as "three factorial." What appears to be an exclamation point after the 3 is really a **factorial symbol (!).** The factorial symbol indicates that we are to find the product of decreasing positive integers.

In Excel we can locate the **factorial function** using the same method we used in Section 4 -3 for generating a random number.

1) **Click on Insert**, highlight **Function** and click. The **Insert Function Dialog** box will open.

2) **Highlight Math & Trig**

3) Scroll through the function names and **highlight FACT**.

4) Click **OK**

5) In the FACT dialog box **enter that number you wish to expand by using factorials**.

    a. FACT(3) returns a result of 6.

    b. FACT(5) returns a result of 120

6) Click **OK**

Note: Remember that FACT(5) is equivalent to $5 \cdot 4 \cdot 3 \cdot 2 \cdot 1$

SECTION 4-6: PERMUTATIONS AND COMBINATIONS:

Problems involving permutations and combinations such as those found in section 4-7 of your textbook can be done fairly quickly and quite easily with Excel. As noted in your textbook, when using the term permutation of any set or subset of objects or events it is implied that order is taken into account and different orderings of the same items are counted separately. Permutations are different from combinations, which do not count different arrangements separately.

The **PERMUT (permutations)** function can be found in the Function Dialog box in the **Statistical** function category.

The **COMBIN (combinations)** function can be found in the Function Dialog box in the **Math and Trig** function category.

Consider the **Clinical Trial of New Drug example** found in Section 4-7 of your textbook. We will use the **PERMUT function dialog** box to mirror the work done in that example:

Number – an integer that refers to the total number of objects - in this case 10 different subjects are available.

Number_chosen - an integer that identifies the number of objects in each permutation. – in this case 8 subjects are chosen without replacement.

The result is 1,814,400 different possible arrangements of 8 subjects selected from the 10 that are available.

Consider the **Phase I of a Clinical Trial example** found in Section 4-7 of your textbook. In this example it is important to notice that we use are using permutations for part (a) of this example because order does count. We are using combinations for part (b) of this problem because order does not count. We have already addressed part (a) in our work with the PERMUT function. We will focus on part (b) and use the **COMBIN function dialog** box to mirror the work done in the second part of this example:

Number - the number of items – in this case there are 10 available people.

Number_chosen - the number of items in each combination – in this case there are 8 people selected.

There are 45 combinations.

Function Arguments [?] [X]

COMBIN

Number | 10 | = 10

Number_chosen | 8 | = 8

= 45

Returns the number of combinations for a given number of items.

Number_chosen is the number of items in each combination.

Formula result =   45

Help on this function   [OK]   [Cancel]

When all of the different possible arrangements are taken into account there are 1,814,400 permutations. When order is not taken into account there are only 45 combinations.

TO PRACTICE THESE SKILLS

You can practice the Excel skills learned in this section of this manual by working through the problems 18, 19, 21, and 28 from the Basic Skills and Concepts for Section 4-7.

CHAPTER 5: DISCRETE PROBABILITY DISTRIBUTIONS

SECTION 5-1: OVERVIEW **65**

SECTION 5-2: RANDOM VARIABLES AND PROBABILITY DISTRIBUTIONS **65**

CREATING A PROBABILITY HISTOGRAM **65**

COMPUTING THE MEAN, VARIANCE AND STANDARD DEVIATION **67**

IDENTIFYING UNUSUAL RESULTS WITH THE RANGE RULE OF THUMB **68**

EXPECTED VALUE **68**

TO PRACTICE THESE SKILLS **69**

SECTION 5-3: BINOMIAL PROBABILITY DISTRIBUTIONS **69**

CREATING A COMPLETE DISTRIBUTION **69**

CREATING A WORKSHEET THAT WILL COMPUTE JUST A SPECIFIC PROBABILITY **71**

TO PRACTICE THESE SKILLS **72**

SECTION 5-4: MEAN, VARIANCE, AND STANDARD DEVIATION FOR BINOMIAL DISTRIBUTION **72**

TO PRACTICE THESE SKILLS **73**

SECTION 5-5: CREATING A POISSON DISTRIBUTION **74**

TO PRACTICE THESE SKILLS **75**

SECTION 5-1: OVERVIEW

Excel has functions built into it that can be used to calculate the probabilities associated with several different probability distributions. Computing these probabilities by hand can be very time consuming. Although tables are available for some distributions, these are also limited in scope. Excel provides you with a tremendous amount of flexibility in creating these distributions quickly and efficiently. The following list contains an overview of the new functions that will be introduced within this chapter.

FILL SERIES
This feature enables us to quickly and efficiently enter a string of consecutive numbers in a column or row.

BINOMDIST
This function returns the individual term binomial distribution probability.

POISSON
This function returns the Poisson probability that a particular number of occurrences of an event will occur over some interval.

SECTION 5-2: RANDOM VARIABLES AND PROBABILITY DISTRIBUTIONS

Let's suppose that a study consists of randomly selecting 14 newborn babies and counting the number of girls. If we assume that boys and girls are equally likely, and we let x = number of girls among 14 babies, we can look at the probability distribution shown. We will work with the values given, and create a probability histogram, as well as consider how we can use Excel to compute the mean and standard deviation of this probability distribution. In the next section, we will actually learn how to generate the probabilities in this distribution using Excel.

	A	B	C
1	x (girls)	P(x)	
2	0	0.000	
3	1	0.001	
4	2	0.006	
5	3	0.022	
6	4	0.061	
7	5	0.122	
8	6	0.183	
9	7	0.209	
10	8	0.183	
11	9	0.122	
12	10	0.061	
13	11	0.022	
14	12	0.006	
15	13	0.001	
16	14	0.000	
17			
18			

CREATING A PROBABILITY HISTOGRAM

1) **Open a new worksheet.** Again, it is a good idea to get in the habit of entering the data in a worksheet entitled "Original Data." You can then copy the data to another worksheet, and work with the data there, ensuring that you always have ready access to the original data if needed.

2) **Create the Column Headings:** Type "x" in cell A1 of a new worksheet, and "P(x)" in cell B1. Then enter the values given in the table shown starting in cells A2 and B2 respectively.

3) **Copy your data and rename your worksheet:** Select this data, and copy it to another worksheet. Give your new worksheet an appropriate name that will remind you what is there. You might want to call the worksheet something like "prob dist." to indicate that you have the probability distribution in this sheet.

4) **Access the Chart Dialog Box:** Click on the **Chart Icon**, or click on **Insert** in the menu bar, then click on **Chart.**

5) **Choose the Column Chart Type:** Click on **Column** under **Chart Type** and the first option under **Chart sub-type.** Then click on **Next**.

6) **Enter the Data Range:** With your cursor positioned in the dialog box by **Data Range**, select the cells containing your probabilities. In the worksheet we set up, this would be cells B2 through B16. Notice that the box will show ='prob dist'!$B$2:$B$16 if you named your sheet "prob dist."

7) **Indicate data is in columns:** Make sure that the bubble in front of **Columns** is selected.

8) **Enter the horizontal axis values:** Click on the **Series** tab, and position your cursor in the dialog box by the word **Category (X) axis labels.** Select the cells containing the x values. In the worksheet we set up, this would be cells A2 through A16. Notice that the box will show ='prob dist'!$A$2:$A$16 if you named your sheet "prob dist."

9) **Name your graph and axes:** Click on **Next**, and give your graph and your axes appropriate names. Then click on **Finish**.

10) **Modify your graph:** You will need to modify your graph. You can click on the Legend box, and delete this. You should double click on one of the bars, and in the **Format Data Series** dialog box, click on **Options**, and set your gap width to 0. You will probably also want to resize your chart area. If you need more specific instructions, look back in Chapter 2, and follow the appropriate instructions in section 2-3. When you are done, your probability distribution should look similar to the one below.

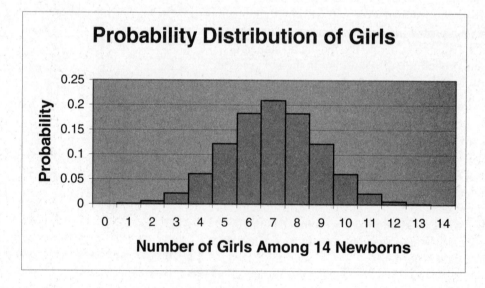

COMPUTING THE MEAN, VARIANCE AND STANDARD DEVIATION

We can compute the mean, variance and standard deviation of a probability distribution according to formulas 5 – 1, 5 – 3 and 5 – 4 in your book. To use Excel to work with these formulas, follow the steps below. The work below assumes that your x values are in column A, starting in cell A2, and that your probabilities are in column B, starting in cell B2. (See the print out on the next page.) If your data is in a different location, you should make adjustments as necessary to the specific cell information shown below.

1) **Create the products of random variables and corresponding probabilities:** In cell D1, type in: x * P(x) as the title for your column. Then in cell D2, type in the following formula: = A2 * B2. This tells Excel to multiply the random variable in cell A2 by its associated probability in cell B2. After pressing **Enter**, you should see a 0 in cell D2, since both values being multiplied are 0.

2) **Copy your formula down:** Position your cursor back in cell D2. Then use the fill handle to copy the formula down through cell D16.

3) **Find the mean:** Formula 5-1 tells us that we need to add these products up. Position your cursor in cell C18, and type "Mean". Then position your cursor in cell D18, and click on the **Summation** icon in the toolbar. To find the sum of the appropriate values, select the cells D2 to D16. Notice that the formula: =SUM(D2:D16) appears in cell D18, and the column of values above this cell are selected. Press **Enter.** You should now see the value 6.993 in cell D 18. This is the mean of the probability distribution.

4) **Find the Variance:** To find the variance, we will use Formula 5-3. We need to create another column which allows us to compute $x^2 \cdot P(x)$.

   a. Position your cursor in cell E1, and type in the formula: x^2*P(x).

   b. Move your cursor to cell E2, and type in the formula: = A2^2*B2. Then press **Enter**. This formula takes the value in cell A2, squares it, and then multiplies it by the value in cell B2.

   c. Reposition your cursor in cell E2, and use the fill handle to fill the column down through E16.

   d. Position your cursor in cell E17, and click on the summation icon in the toolbar. You should see the formula: =SUM(E2:E16) appear in cell E17. Press **Enter**, and you will see the value 52.467.

   e. Position your cursor in cell C19 and type in the word "Variance". Then move to D19 and type in the formula: =$E$17-$D$18^2. This will take the sum of your products and subtract the square of the mean from this sum. This corresponds to Formula 5-3 found in your book. You may want to decrease the number of decimal places that initially appear in your answer.

5) **Find the Standard Deviation:** Now position your cursor in cell C20, and type in "Std. Dev." Move to cell D 20 and type the following formula: =SQRT($D$19). This will take the square root of the variance, which will produce your standard deviation. Your worksheet should look similar to the one shown below:

	A	B	C	D	E	F
1	x (girls)	P(x)		x* P(x)	x^2*P(x)	
2	0	0.000		0.000	0	
3	1	0.001		0.001	0.001	
4	2	0.006		0.012	0.024	
5	3	0.022		0.066	0.198	
6	4	0.061		0.244	0.976	
7	5	0.122		0.610	3.05	
8	6	0.183		1.098	6.588	
9	7	0.209		1.463	10.241	
10	8	0.183		1.464	11.712	
11	9	0.122		1.098	9.882	
12	10	0.061		0.610	6.1	
13	11	0.022		0.242	2.662	
14	12	0.006		0.072	0.864	
15	13	0.001		0.013	0.169	
16	14	0.000		0.000	0	
17					52.467	
18			Mean	6.993		
19			Variance	3.564951		
20			Std. Dev.	1.888108		
21						
22						

IDENTIFYING UNUSUAL RESULTS WITH THE RANGE RULE OF THUMB

The range rule of thumb tells us that most values should lie within 2 standard deviations of the mean. I we have computed the mean and standard deviation in Excel, we can easily create formulas which allow us to determine the minimum and maximum usual values.

1) **Creating the minimum usual value:** In the worksheet from above, type the word Minimum in cell C22. Move to cell D22 and type in the following formula: = d18-2*d19. This formula subtracts 2 times the standard deviation from the mean. Press **Enter** to see the result.

2) **Creating the maximum usual value:** Type the word Maximum in cell C23. Move to cell D23 and type in the following formula: =d18+2*d19. This formula adds 2 times the standard deviation to the mean. Press **Enter** to see the result. Because of the context of this problem, we would interpret this to mean that the usual number of girls in a randomly selected group of 14 newborns would be between 0 and 11.

22			Minimum	-0.136902	
23			Maximum	10.76921557	
24					

EXPECTED VALUE

Notice that the value that you computed in your worksheet for the summation of the products of your random variables and their corresponding probabilities can also be called the expected value of a discrete random variable.

TO PRACTICE THESE SKILLS

You can apply these technology skills by working on exercises 7, 8, 9 and 10 from section 5–2 Basic Skills and Concepts in your textbook.

SECTION 5-3: BINOMIAL PROBABILITY DISTRIBUTIONS

Binomial variables take on only two values. One of these values is generally designated as a "success" and the other a "failure." We typically see the probability of a "success" denoted by the letter p and the probability of failure denoted by the letter q. The sum of p and q must equal one, since success or failure are the only possible outcomes.

Suppose we consider a binomial distribution where p = .65 and there are 15 trials. Since there are 15 trials, we know that the random variable can take on the values between 0 and 15 inclusive.

CREATING A COMPLETE DISTRIBUTION

Begin your work in a new worksheet. You may want to name this worksheet something like "Bin. Dist" to indicate that you are creating a binomial distribution.

1) **Create the column of random variables:** First enter a title for the column representing the random variable in cell A1. (If we knew what the random variable represented, we should use a suitable name, otherwise, we will use "x".) Then in cell A2, type in the value 0, since this is the first value of the random variable. Press **Enter**.

2) **Use the Fill Series feature:** Reposition your cursor in cell A2. (Note: In order to activate the fill feature, you **must** move out of the cell where your first value is typed, and then move back to it.) From the command bar, click on **Edit, Fill,** and then click on **Series.** You will see the **Series** dialog box open.

   a. Make sure that the bubble in front of **Columns** is checked.

   b. Make sure that **Linear** is selected under **Type**.

   c. Make sure that the **Step value** is set at 1.

   d. Type in 15 for the **Stop value**, since there are 15 trials in the experiment.

   e. Click on **OK**. You should now see the whole numbers from 0 to 15 in column A.

3) **Create the column for probabilities:** Move to column B and type "P(x)" in cell B1. Press **Enter**. Your cursor should now be in cell B2.

4) **Access the Binomial Distribution Function:** In the command bar, click on **Insert**, and then click on **Function**. (Notice the symbol preceding **Function** can also be found on the menu bar. You may activate this dialog box by clicking on this icon in the menu bar.) In the "Search for function:" box, type in "binomial distribution". You will see the function **BINOMDIST** highlighted in the "Select a function:" box. Click on **OK.**

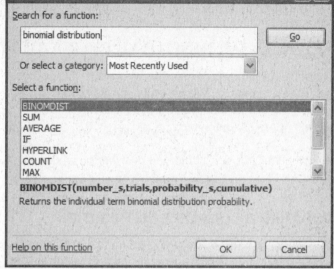

5) **Fill in the BINOMDIST Dialog Box:** You need to complete the dialog box as follows:

   a. **Number_s** refers to the number of successes. You want to enter the cell address where this information is stored. Since we began our values for x in cell A2, we type in A2.

   b. **Trials** refers to the total number of trials. Type in 15.

   c. **Probability_s** refers to the probability of a success. For this experiment, type in the value .65.

   d. **Cumulative** will list the cumulative probabilities. Since we do not want these at this time, type in "False".

   e. Click on **OK.** You will now see the probability of getting exactly 0 successes in 15 trials if the probability or success is .65. This value may be written in exponential notation. You can reformat the column to show the number with more decimal places.

	A	B	C
1	x	P(x)	
2	0	0.0000001449	
3	1	0.0000040361	
4	2	0.0000524687	
5	3	0.0004222484	
6	4	0.0023525267	
7	5	0.0096117520	
8	6	0.0297506611	
9	7	0.0710372928	
10	8	0.1319264010	
11	9	0.1905603570	
12	10	0.2123386835	
13	11	0.1792469406	
14	12	0.1109623918	
15	13	0.0475553108	
16	14	0.0126167151	
17	15	0.0015620695	
18			

6) **Use the fill handle to copy the formula:** With cell B2 activated, you can use the fill handle to fill in the remaining values in the table.

CREATING A WORKSHEET THAT WILL COMPUTE JUST A SPECIFIC PROBABILITY

You may not need to find the entire probability distribution, but may just want to create a worksheet which would make it easy to plug in appropriate values and find a particular probability. Follow the steps below to create such a worksheet.

1) **Give your worksheet a title:** Create a clear title for your worksheet. Notice that we called our worksheet "Worksheet to Compute Specific Binomial Probabilities". Since this name took up more than one cell, it's a good idea to select the cells that the title spans and then click on **Format,** click on **Cells**, then click on the tab that says **Alignment,** and click in the box before **Merge cells.** Then click on **OK.**

2) **Set up the categories you need to input:** Type the specifics of the information that you would need to enter into Excel. You need to enter the values for n, p and x.

3) **Format the background for the cells where you will be typing in values:** To make it clear where you will be typing in your values, you may find it helpful to format those cells with some background color. Position your cursor in a cell where you will need to enter a value. Then click on **Format,** click on **Cells,** and click on the tab that says **Patterns.** Choose the color you want for the background fill in the cell, and then click **OK.** You can then copy this formatting to the other cells by pressing and holding the **Ctrl** key followed by the **C** key on your keyboard. Move your cursor to the next cell you want to have this background color and press and hold the **Ctrl** key followed by the **V** key on your keyboard. To turn off the copy command (notice that you see a blinking frame around the cell you copied), press **Esc.**

4) **In put specific values in the cells:** We will let n = 6, p = .75 and x = 4.

5) **Apply the BINOMDIST function to these values:** In the command bar, click on **Insert**, and then click on **Function**. (Notice the symbol preceding **Function** can also be found on the menu bar. You may activate this dialog box by clicking on this icon in the menu bar.) In the "Search for function:" box, type in "binomial distribution". You will see the function **BINOMDIST** highlighted in the "Select a function:" box. Click on **OK.** Fill in the dialog box using cell references. For the worksheet we set up, we would fill our dialog box out as the one shown. Then click on **OK.**

Function Arguments

BINOMDIST

Number_s B7 = 4
Trials B3 = 6
Probability_s B5 = 0.75
Cumulative false = FALSE

= 0.296630859

Returns the individual term binomial distribution probability.

Cumulative is a logical value: for the cumulative distribution function, use TRUE; for the probability mass function, use FALSE.

Formula result = 0.296630859

Help on this function

6) **Your final worksheet:** Your finished worksheet should look like the one shown.

	A	B	C
1	Worksheet to compute specific binomial probabilities		
2			
3	Enter the sample size (n) :	6	
4			
5	Enter the probability of success (p):	0.75	
6			
7	Enter the number of successes (x):	4	
8			
9	The value of P(x) is:	0.296631	
10			
11			
12			

TO PRACTICE THESE SKILLS

You can apply these technology skills by working on the following exercises. Make sure you save your work using a file name that is indicative of the material contained in your worksheets.

1) Read exercise 29 from Section 5-3 Basic Skills and Concepts in your textbook. Use Excel to set up the probability distribution for this exercise. Before beginning your work in Excel, make sure you clearly identify what the values for your random are, and what value you should use for the probability of "success." You can use Excel to compute the sums of appropriate cells from your probability distribution. For example, to find the probability of at least one household, you can add up the probabilities for x = 1 to x = 10.

2) Read exercise 30 from Section 5-3 Basic Skills and Concepts in your textbook. Use Excel to set up the probability distribution for this exercise. Use your table to answer the questions asked.

3) You can use the worksheet you created for computing specific binomial probabilities for exercises 15 through 24 from section 5-3 Basic Skills and Concepts in your textbook.

SECTION 5-4: MEAN, VARIANCE, AND STANDARD DEVIATION FOR BINOMIAL DISTRIBUTION

Although you can use the formulas presented in section 5-2 to compute the mean, variance and standard deviation for the binomial distribution, there are easier formulas to work with for this particular distribution.

- You can find the mean by multiplying the sample size (n) by the probability of success (p).

- You can find the standard deviation by taking the square root of the product formed by multiplying the sample size (n) by the probability of success (p) and the probability of failure (q).

- You should use formulas preceded by the "=" sign when you enter your information into Excel.

Suppose we used the values: n = 14, p = 0.5 and q = 0.5. You can follow the steps below to create a table for this binomial experiment.

1) **Give your worksheet a title:** Create a clear title for your worksheet. Notice that we called our worksheet "Mean and Standard Deviation for a Binomial Probability". Since this name took up

more than one cell, it's a good idea to select the cells that the title spans and then click on **Format,** click on **Cells,** then click on the tab that says **Alignment,** and click in the box before **Merge cells.** Then click on **OK.**

2) **Set up the categories you need to input:** In our worksheet, we typed the information "Sample Size", "Probability of Success", "Probability of Failure", "Mean", and "Std. Dev" in cells A2 through A6.

3) **Format the background for the cells where you will be typing in values:** To make it clear where you will be typing in your values, you may find it helpful to format those cells with some background color. Position your cursor in a cell where you will need to enter a value. Then click on **Format,** click on **Cells,** and click on the tab that says **Patterns.** Choose the color you want for the background fill in the cell, and then click **OK.** You can then copy this formatting to the other cells by pressing and holding the **Ctrl** key followed by the **C** key on your keyboard. Move your cursor to the next cell you want to have this background color and press and hold the **Ctrl** key followed by the **V** key on your keyboard. To turn off the copy command (notice that you see a blinking frame around the cell you copied), press **Esc.**

4) **Type in your specific values:** We will use an example where n is 14, and p is .5. In cell B2 type in 14. In cells B3 type in 0.5.

5) **Set up a formula to compute q:** Since q can be computed by subtracting p from 1, in cell B4, type in =1-B3. When you press **Enter**, you should see the value shown in the table below.

6) **Set up the formula to compute the mean:** Since the mean of a binomial probability distribution is n * p, in cell B5, type in the formula: =b2*b3. Then press **Enter.**

7) **Set up the formula to compute the standard deviation:** Since the standard deviation of a binomial probability distribution is $\sqrt{n \cdot p \cdot q}$, in cell B6 type in the formula: =sqrt(b2*b3*b4). Then press **Enter.**

8) **Your final table:** Your table should look similar to the one shown below:

	A	B	C
1	**Mean and Standard Deviation for a Binomial Probability**		
2	Sample Size (n)	14	
3	Probability of Success (p)	0.5	
4	Probability of Failure (q)	0.5	
5	Mean	7	
6	Std. Dev	1.870828693	
7			

TO PRACTICE THESE SKILLS

Once you have the table from above set up in a worksheet, you can change the values for sample size, probability of success and probability of failure. Your values for the mean and standard deviation should be automatically updated. You can copy this table, paste it in other cells of your worksheet, and then change the numbers in the copy of the table for a different problem.

1) Create a table similar to the one above, but using the data from exercise 9 from Section 5-4 Basic Skills and Concepts in your textbook. To complete part b, you want to compute the values that are 2 standard deviations above and below the mean. Add lines to your table which will use the numbers generated to compute the minimum usual value ($\mu - 2\sigma$) and the maximum usual value ($\mu + 2\sigma$). You may find it helpful to refer back to section 5-2 of this manual under Identifying Unusual Results with the Range Rule of Thumb.

2) You can now make additional copies of your table, and paste them to other parts of your worksheet if you want to keep a separate record for each problem, or you can merely update the information in the one table, and just keep track of your answers on separate paper. You can work on exercises 10 through 20 from Section 5-4 Basic Skills and Concepts in your textbook.

SECTION 5-5: CREATING A POISSON DISTRIBUTION

In a Poisson distribution, the random variable x is the number of occurrences of the event in an interval. The interval can be time, distance, area, volume, or some similar unit.

Using the example on **World War II Bombs** from section 5-5 in your text, we can generate a table similar to Table 54 in your book, showing the probabilities that a region was hit 0, 1, 2, 3, 4 or 5 times. In this example, the computed mean is 0.929.

1) **Create a column for the random variable:** In cell A1, type in "x". Then type the values 0, 1, 2, 3, 4 and 5 in that column. (You could use the **Edit, Fill, Series** sequence described in section 5-3 to accomplish this also.)

2) **Create a column for the probabilities:** In cell B1, type in "P(x)".

3) **Access the POISSON Function:** Position your cursor in cell B2. Then click on **Insert**, and then click on **Function**. (Notice the symbol preceding **Function** can also be found on the menu bar. You may activate this dialog box by clicking on this icon in the menu bar.) In the "Search for function:" box, type in "poisson". You will see the function **POISSON** highlighted in the "Select a function:" box. Click on **OK**.

4) **Fill in the POISSON Dialog Box:** You should fill in the dialog box as follows:

   a. For the **X** input, type in cell A2, since this is where your first random variable is located.

   b. In the input box following **Mean**, type in .929 since this is the computed mean for this example.

   c. In the **Cumulative** box, type in "**False**".

   d. Click on **OK**. You will now see the probability that a randomly selected region with an area of .25 square kilometers was hit exactly zero times. To match the results in the book, you can format your column to show 3 decimal places.

5) **Use the fill handle to fill in the remaining cells:** With your cursor on cell B2, move your mouse until you see the black plus sign in the lower right hand corner. Holding your left mouse button down, drag this down to fill the rest of the column.

6) **Set up a column for the Expected Number of Regions:** Suppose we want to use this probability to compute the Expected Number of Regions as found in Table 54 in your textbook. In cell C1, type in "Expected Number of Regions". To have the text "wrap" to fit in to the formatted cell width, click on **Format, Cell,** then click on the **Alignment** tab, and click on **Wrap Text**.

7) **Compute the Expected Number of Regions:** Move to cell C2. We want to multiply each probability by the total number of regions (576). Enter the formula: =576*b2 into cell C2. Then use the fill handle to copy this formula down into the remaining cells. Again, to match the style of the table in the book, you can format your column to show 1 decimal place. You will notice that some of the numbers generated are slightly different from those in the table in the book. This is because even when Excel is displaying only 3 decimal digits based on our cell formatting, it is using the longer string in computations.

	A	B	C	D
1	x	P(x)	Expected Number of Regions	
2	0	0.395	227.5	
3	1	0.367	211.3	
4	2	0.170	98.2	
5	3	0.053	30.4	
6	4	0.012	7.1	
7	5	0.002	1.3	
8				
9				

8) **Finishing touches:** To make your table easier to read, you can select the entire table, and then press the centering icon in your tool bar ▤. Your table should now look like the one shown.

TO PRACTICE THESE SKILLS

You can apply these technology skills by working on the following exercises. Make sure you save your work using a file name that is indicative of the material contained in your worksheets.

1) Create a table showing "x" and "P(x)" for exercise 9 from Section 5-5 Basic Skills and Concepts in your textbook.

2) Create a table showing "x", "P(x)" and the "Expected Number" columns for exercise 10 from Section 5-5 Basic Skills and Concepts in your textbook.

CHAPTER 6: NORMAL PROBABILITY DISTRIBUTIONS

SECTION 6-1: OVERVIEW 77

SECTION 6-2: WORKING WITH THE STANDARD NORMAL DISTRIBUTION 77

FINDING PROBABILITIES GIVEN VALUES 77

FINDING A SCORE WHEN GIVEN THE PROBABILITY 80

TO PRACTICE THESE SKILLS 82

SECTION 6-3: APPLICATIONS OF NORMAL DISTRIBUTIONS 82

FINDING PROBABILITIES USING THE NORMDIST FUNCTION 82

FINDING VALUES FROM KNOWN AREAS 85

TO PRACTICE THESE SKILLS 85

SECTION 6-4: SAMPLING DISTRIBUTIONS AND ESTIMATORS 86

SECTION 6-5: THE CENTRAL LIMIT THEOREM 86

VISUALIZING THE CENTRAL LIMIT THEOREM 86

COMPUTING PROBABILITIES THAT INVOLVE THE CENTRAL LIMIT THEOREM 89

TO PRACTICE THESE SKILLS 90

SECTION 6-6: NORMAL DISTRIBUTION AS APPROXIMATION TO BINOMIAL 91

TO PRACTICE THESE SKILLS 94

SECTION 6-7: DETERMINING NORMALITY 94

TO PRACTICE THESE SKILLS 95

SECTION 6-1: OVERVIEW

In this chapter we will explore how to compute probabilities for a normal distribution, as well as find specific values if we are given information about the probability for a particular normal distribution. There are essentially five different functions that we can use when exploring normal distributions.

NORMSDIST: This function returns the **standard** normal cumulative distribution for a specified z value.

NORMSINV: This function returns the inverse of the standard normal cumulative distribution for a specified z value.

STANDARDIZE: This function returns a standardized score value for specified values of the random variable, mean and standard deviation.

NORMDIST: This function returns the cumulative normal distribution for specified values of the random variable, mean and standard deviation.

NORMINV: This function returns the inverse of the normal cumulative distribution for specified values for the probability, mean and standard deviation.

SECTION 6-2: WORKING WITH THE STANDARD NORMAL DISTRIBUTION

The first normal distribution presented in your text is the standard normal distribution. This distribution has a mean of 0 and a standard deviation of 1. In this section of the manual, we will set up worksheets which will allow us to enter information for the different situations that might arise for computing probabilities and finding values. We can then simply enter values into appropriate cells of the worksheets to find the answers requested.

FINDING PROBABILITIES GIVEN VALUES

In order to use Excel to find probabilities, we must recognize that the program computes probabilities by determining the total area under the normal distribution **from the left up to a vertical line at a specific value**. This is vital for us to remember as we work to set up appropriate formulas.

1) **Give your worksheet a title:** Create a clear title for your worksheet. Notice that we called our worksheet "Worksheet to Compute Probabilities for the Standard Normal Distribution". Since this name took up more than one cell, it's a good idea to select the cells that the title spans and then click on **Format,** click on **Cells,** then click on the tab that says **Alignment,** and click in the box before **Merge cells.** Then click on **OK.**

2) **Set up the categories you need to input:** We need to create a worksheet which would allow us to enter in one boundary value, and then compute either P (z < a) or P (z > a). We also need to set up a situation where we could enter in a lower and upper value and compute P (a < z < b). In order to find this probability, we need to generate the values P (z < b) and P (z < a) in Excel. Use the worksheet shown below to set up appropriate categories.

3) **Format the background for the cells where you will be typing in values:** To make it clear where you will be typing in your values, you may find it helpful to format those cells with some background color. Position your cursor in a cell where you will need to enter a value. Then click on **Format,** click on **Cells,** and click on the tab that says **Patterns.** Choose the color you want for the

background fill in the cell, and then click **OK**. You can then copy this formatting to the other cells by pressing and holding the **Ctrl** key followed by the **C** key on your keyboard. Move your cursor to the next cell you want to have this background color and press and hold the **Ctrl** key followed by the **V** key on your keyboard. To turn off the copy command (notice that you see a blinking frame around the cell you copied), press **Esc.**

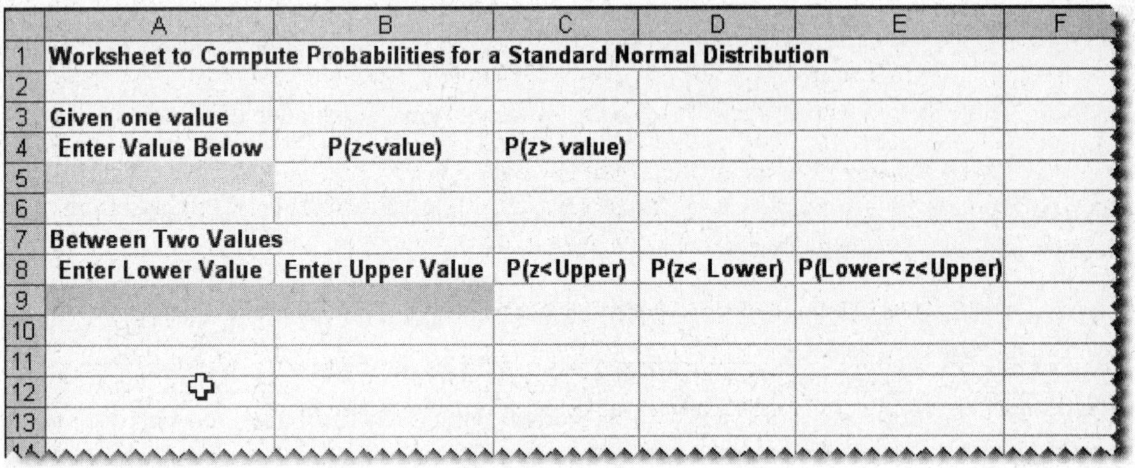

	A	B	C	D	E	F
1	Worksheet to Compute Probabilities for a Standard Normal Distribution					
2						
3	Given one value					
4	Enter Value Below	P(z<value)	P(z> value)			
5						
6						
7	Between Two Values					
8	Enter Lower Value	Enter Upper Value	P(z<Upper)	P(z< Lower)	P(Lower<z<Upper)	
9						
10						
11						
12						
13						

4) **In put specific values in the cells:** Let's use a value of -1.23 for the one value situation, and let's use values of -2 and 1.5 for the two value situation. If you set your worksheet up like the one shown above, you should enter -1.23 in cell A5, -2 in cell A9, and 1.5 in cell B9.

5) **Set up the formula for P (z < value) by accessing the NORMSDIST function:** Remember, Excel will always return a value which represents the area to the left of a particular value. Position your cursor in cell B5 and in the command bar, click on **Insert**, and then click on **Function**. (Notice the symbol preceding **Function** can also be found

Function Arguments

NORMSDIST

Z A5 = -1.23

= 0.109348552

Returns the standard normal cumulative distribution (has a mean of zero and a standard deviation of one).

Z is the value for which you want the distribution.

Formula result = 0.109348552

Help on this function   OK   Cancel

on the menu bar. You may activate this dialog box by clicking on this icon in the menu bar.) In the "Search for function:" box, type in "standard normal distribution". You will see the function **NORMSDIST** highlighted in the "Select a function:" box. Click on **OK**. Fill in the dialog box using the appropriate cell reference for where your value is located. For the worksheet we set up, we would fill our dialog box out as the one shown. Then click on **OK**.

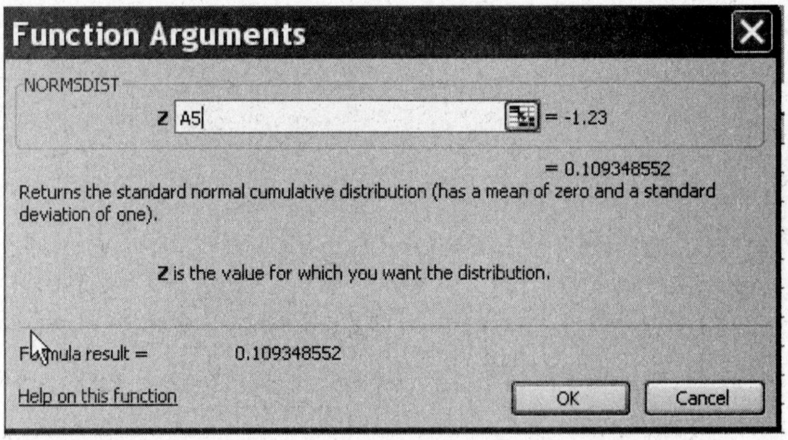

6) **Copy and update this formula to the other cells:** Rather than accessing the dialog box twice more for the remaining cells where you want P (z < value), we can copy the formula we set up in cell B5 to the other cells. **Always make sure that you have appropriate cell references after you have copied the formula to a new location!**

   a. **Position your cursor in cell B5:** Press and hold **Ctrl** and then press **C** on your key board. You should now see a blinking border around B5.

b. **Move your cursor to cell C9:** Press and hold **Ctrl** and then press **V** on your key board. Your formula should have been copied to this new cell.

c. **Check the formula line at the top of the page:** Always make sure you check that when you copied the formula, you are referring to the appropriate cells. In this case, we want P (z < Upper). Since the upper value is in cell B9, the formula is correct.

d. **Move your cursor to cell D9:** Press and hold down **Ctrl** and then press **V** on your keyboard. This will copy the formula again.

e. **Check the formula line at the top of the page.** Notice that you want P (z < Lower), but the formula bar shows cell C9. Position your cursor up in the formula line right in front of the letter C. Delete C and type in a, since your lower value is in cell A9.

f. **Your sheet should look like the one shown:**

	A	B	C	D	E	F
1	Worksheet to Compute Probabilities for a Standard Normal Distribution					
2						
3	Given one value					
4	Enter Value Below	P(z<value)	P(z> value)			
5	-1.23	0.109348552				
6						
7	Between Two Values					
8	Enter Lower Value	Enter Upper Value	P(z<Upper)	P(z< Lower)	P(Lower<z<Upper)	
9	-2	1.5	0.933192799	0.022750132		
10						
11						

7) **Create formulas for the remaining cells:** Since Excel's NORMSDIST function will only return the area under the normal curve to the left of the value we use, we need to create formulas that use this information to help determine P (z > value) and P (Lower < z < Upper).

a. **P (z > value):** Since the area under the entire normal curve is 1, if we want P (z > value), we can use the formula 1 – P (z < value). Position your cursor in cell C5, and type in the formula =1-b5. Then press **Enter**. You should now see 0.890651448 in cell C5.

b. **P(Lower < z < Upper):** Since the area we want is between the upper and lower value, we can create the formula which computes P (z < Upper) – P (z < Lower). Position your cursor in cell E9 and type in the formula =c9-d9. Then press **Enter.** You should now see 0.910442667 in cell E9.

8) **Try some other values:** You can now just type in other values in the shaded cells, and your spreadsheet will automatically update the values for the probabilities. Check that your worksheet is set up correctly by typing the values shown in the yellow cells in the worksheet below. Your values should be updated to show the values for the probabilities below.

	A	B	C	D	E	F
1	Worksheet to Compute Probabilities for a Standard Normal Distribution					
2						
3	Given one value					
4	Enter Value Below	P(z<value)	P(z> value)			
5	1.25	0.894350226	0.105649774			
6						
7	Between Two Values					
8	Enter Lower Value	Enter Upper Value	P(z<Upper)	P(z< Lower)	P(Lower<z<Upper)	
9	1	2	0.977249868	0.841344746	0.135905122	
10						

FINDING A SCORE WHEN GIVEN THE PROBABILITY

Let's assume that we are working with thermometers that are normally distributed with a mean of 0 degrees Celsius and a standard deviation of 1 degree Celsius. Suppose we are given a probability and we want to find the value that corresponds to that probability.   A classic example of this would be if we wanted to find the value for a particular percentile. Remember P_{80} represents the value such that at least 80% of the values would be less than or equal to that value and at least 20% of the values were greater than or equal to that value.

Again, let's work to set up a worksheet which will allow us to enter appropriate values into certain cells, and which then would automatically compute the values we wanted to know. We want to create a worksheet that looks like the one below.

	A	B	C	D	E	F
1	Finding Values from Known Areas					
2	If given the area to the LEFT of the value	Enter area as decimal	Enter Mean	Enter SD	Value to 2 decimal places	
3		0.8	0	1	0.84	
4						
5	If given the area to the RIGHT of the value	Enter area as decimal	Enter Mean	Enter SD	Value to 2 decimal places	
6		0.2	0	1	0.84	
7						
8						

1) **Set up your column headings:** Set up a chart like the one above, making use of the **Format, Cells, Alignment,** and **Wrap Text** feature to have the titles fit as shown.

2) **Format the background color for the cells where we will enter information:** Again, there are certain cells where we will enter information for the situation. While we know the mean and the standard deviation of a standard normal distribution remain at 0 and 1 respectively, we are creating a sheet that can also be used for other normal distributions. Notice that two situations are set up: if you're given the area to the **Left** of the value you want, and if you're given the area to the **Right** of the value you want.

3) **Access the NORMINV Function to find the desired value where area is to the LEFT of the value:**

   a. **Position your cursor in cell E3**, and click on the **Function** icon on the toolbar (or click on **Insert, Function**). In the "Search for function:" box, type in "inverse of normal distribution".   You will see the function **NORMINV**

Insert Function

Search for a function:

inverse of normal distribution   Go

Or select a category: Recommended

Select a function:

NORMINV
NORMSINV
NORMSDIST
LOGINV
TINV

NORMINV(probability,mean,standard_dev)
Returns the inverse of the normal cumulative distribution for the specified mean and standard deviation.

Help on this function   OK   Cancel

highlighted in the "Select a function:" box. Click on **OK.**

b. **Fill in the dialog box**: You will use the cell reference where the area is located for **Probability.** You should also use the appropriate cell references for where the **Mean** and **Standard Deviation** are found. If you set your worksheet up like the one shown, your dialog box should look like the one shown. **Keep in mind that Excel will always interpret the probability you put in as being to the LEFT of the value you want.**

Function Arguments

NORMINV
Probability B3 = 0.8
Mean C3 = 0
Standard_dev D3 = 1

= 0.841621234

Returns the inverse of the normal cumulative distribution for the specified mean and standard deviation.

Standard_dev is the standard deviation of the distribution, a positive number.

Formula result =   0.84

Help on this function   OK   Cancel

4) **Find the value where area is to the RIGHT of the value:** We can again copy the formula we have entered above, and paste it into the cell which represents the value for area to the **RIGHT.** However, we will have to modify the formula.

a. **Position your cursor in cell E3:** Press and hold **Ctrl** and then press **C** on your key board. You should now see a blinking border around E3.

b. **Move your cursor to cell E6:** Press and hold **Ctrl** and then press **V** on your key board. Your formula should have been copied to this new cell.

c. **Check the formula line at the top of the page:** Always make sure you check that when you copied the formula, you are using appropriate values. Keeping in mind that Excel always understands that Probability is the area to the LEFT of the value you want, we need to make a minor change to our copied formula. Your copied formula should look like: =NORMINV(B6,C6,D6). We need to realize that since cell B6 contains the area (or probability) to the RIGHT of the value we want, we need to adjust the formula so that we see =NORMINV(1-B6,C6,D6). Position your cursor in the formula line at the top of the worksheet right before the B, and type in 1 - . Then press **Enter.** You should now see .84 in cell E6. This should make sense. If the value .84 separates the bottom 80% of values, it should be the same score which separates the top 20% of values.

5) **Try some other values:** To make sure that your worksheet is set up correctly, enter an area to the **LEFT** of .025, and an area to the **RIGHT** of .025. Keep your mean and standard deviation at 0 and 1 respectively.

	A	B	C	D	E	F
1	Finding Values from Known Areas					
2	If given the area to the LEFT of the value	Enter area as decimal	Enter Mean	Enter SD	Value to 2 decimal places	
3		0.025	0	1	-1.96	
4						
5	If given the area to the RIGHT of the value	Enter area as decimal	Enter Mean	Enter SD	Value to 2 decimal places	
6		0.025	0	1	1.96	
7						
8						

The values shown in the table should make sense. If you have an area of .025 to the LEFT of the value you want, the value will have to be to the left of the mean, and since this is the standard normal distribution, it must therefore be negative. If you have an area of .025 to the RIGHT of the value you want, the value will have to be to the right of the mean. Since this is the standard normal distribution, the value must therefore be positive. Since you have equal areas in the "tails", it makes sense that the values are opposites of each other for the standard normal distribution.

TO PRACTICE THESE SKILLS

1) You can apply the skills you learned in this section to find probabilities by using Excel to complete exercises 9 through 26 and 33 through 36 from Section 6-2 Basic Skills and Concepts in your textbook.

2) You can apply the skills you learned in this section to find scores associated with particular percentiles or probabilities by using Excel to complete exercises 37 through 40 from Section 6-2 Basic Skills and Concepts in your textbook.

SECTION 6-3: APPLICATIONS OF NORMAL DISTRIBUTIONS

Your textbook presents the idea that if we are working with a normal distribution which is not a standard normal, we could opt to "standardize" the scores, and then use the techniques for finding probabilities and values as presented in the Standard Normal section.

Since Excel allows us to work directly with non-standard normal distributions, we will modify our previous spreadsheet entitled "Worksheet to Compute Probabilities for the Standard Normal Distribution". (See section 6-2.)

FINDING PROBABILITIES USING THE NORMDIST FUNCTION

If you completed and saved your work from section 6-2, you can modify that worksheet rather than creating an entirely new one. You would need to change the title, and insert a couple of more columns for the additional information we need for a non-standard normal distribution. You will also have to change from the NORMSDIST function to the NORMDIST function. If you feel comfortable with this, simply copy and paste your standard normal distribution worksheet to a new worksheet, and begin making appropriate modifications. You will want to make sure you review the step where we introduce the NORMDIST function. Also, notice that we no longer use "z" to indicate a value, we use "x". The letter "z" is specifically reserved for the standard normal distribution.

You may want to just begin again, so that you can practice setting up worksheets. The directions below take you through each step.

1) **Give your worksheet a title:** Create a clear title for your worksheet. Notice that we called our worksheet "Worksheet to Compute Probabilities for a Normal Distribution". Since this name took up more than one cell, it's a good idea to select the cells that the title spans and then click on **Format,** click on **Cells,** then click on the tab that says **Alignment,** and click in the box before **Merge cells.** Then click on **OK.**

2) **Set up the categories you need to input:** We need to create a worksheet which would allow us to enter in one boundary value, and then compute either P (x < a) or P (x > a). We also need to set up a

situation where we could enter in a lower and upper value and compute $P(a < x < b)$. In order to find this probability, we need to generate the values $P(x < b)$ and $P(x < a)$ in Excel. Use the worksheet shown below to set up appropriate categories.

3) **Format the background for the cells where you will be typing in values:** To make it clear where you will be typing in your values, you may find it helpful to format those cells with some background color. Position your cursor in a cell where you will need to enter a value. Then click on **Format,** click on **Cells,** and click on the tab that says **Patterns.** Choose the color you want for the background fill in the cell, and then click **OK.** You can then copy this formatting to the other cells by pressing and holding the **Ctrl** key followed by the **C** key on your keyboard. Move your cursor to the next cell you want to have this background color and press and hold the **Ctrl** key followed by the **V** key on your keyboard. To turn off the copy command (notice that you see a blinking frame around the cell you copied), press **Esc.**

	A	B	C	D	E	F
1	Worksheet to Compute Probabilities for a Normal Distribution					
2						
3	Given One Value					
4	Enter Value	Enter Mean	Enter SD		P(x < Value)	P(x > Value)
5						
6						
7	Between Two Values					
8	Enter Lower Value	Enter Upper Value	Enter Mean	Enter SD	P(x < Upper)	
9					P(x < Lower)	
10					P(Lower < x < Upper)	
11						
12						
13						

4) **In put specific values in the cells:** Let's use data which assumes that weights of men are normally distributed with a mean of 172 pounds and a standard deviation of 29 pounds. Let's consider the probability that a randomly selected man will weigh less than 174 pounds. Let's also consider the probability that a randomly selected man will weigh between 150 and 170 pounds.

  a. **Enter the One Value:** We want to find the probability that a randomly selected man will weigh less than 174 pounds, so we will enter 174 in cell A5, 172 in cell B5 and 29 in cell C5.

  b. **Enter the Two Values:** We want to find the probability that a randomly selected man will weigh between 150 and 170 pounds. We will enter 150 in cell A9, 170 in B9, 174 in C9 and 29 in D9.

5) **Set up the formula for P (x < value) by accessing the NORMDIST function:**

Insert Function dialog box:

Search for a function: normal distribution [Go]

Or select a category: Recommended

Select a function:
NORMSDIST
NORMDIST
NORMSINV
NORMINV
KURT
ZTEST

NORMDIST(x,mean,standard_dev,cumulative)
Returns the normal cumulative distribution for the specified mean and standard deviation.

Help on this function [OK] [Cancel]

Remember, Excel will always return a value which represents the area to the left of a particular value. Position your cursor in cell E5 and in the command bar, click on **Insert**, and then click on **Function**. (Notice the symbol preceding **Function** can also be found on the menu bar. You may activate this dialog box by clicking on this icon in the menu bar.) In the "Search for function:" box, type in "normal distribution". You will see two normal distribution functions listed. **Make sure you highlight the function NORMDIST in the "Select a function:" box.** Click on **OK**. Fill in the dialog box using the appropriate cell references for where your values are located. For the worksheet we set up, we would fill our dialog box out as the one shown. Then click on **OK**.

Function Arguments

NORMDIST

X	A5	= 174
Mean	B5	= 172
Standard_dev	C5	= 29
Cumulative	true	= TRUE

= 0.527491466

Returns the normal cumulative distribution for the specified mean and standard deviation.

Cumulative is a logical value: for the cumulative distribution function, use TRUE; for the probability mass function, use FALSE.

Formula result =   0.527491466

Help on this function   OK   Cancel

6) **Copy and update this formula to the other cells:** Rather than accessing the dialog box again for the remaining cells where you want P (z < value), we can copy the formula we set up in cell E5 to the other cells. **Always make sure that you have appropriate cell references after you have copied the formula to a new location!**

   a. **Position your cursor in cell E5:** Press and hold **Ctrl** and then press **C** on your key board. You should now see a blinking border around E5.

   b. **Move your cursor to cell F8:** Press and hold **Ctrl** and then press **V** on your key board. Your formula should have been copied to this new cell, however you see a screen that looks like:

P(x < Upper)	#VALUE!
P(x < Lower)	
P(Lower < x < Upper)	

   c. **Check the formula line at the top of the page:** Always make sure you check that when you copied the formula, you are referring to the appropriate cells. In this case, we want P (x < Upper). Notice that the formula at the top of the page shows: =NORMDIST(B8,C8,D8,TRUE). Notice that because of the way we set up our worksheet, we actually want B9, C9 and D9. Simply position your cursor in the formula line and make appropriate changes so that you see: =NORMDIST(B9,C9,D9,TRUE). Now press **ENTER**.

   d. **Move your cursor to cell F9:** Press and hold down **Ctrl** and then press **V** on your keyboard. This will copy the formula again.

   e. **Check the formula line at the top of the page**. Notice that you want P (x < Lower), but the formula bar shows cell B9, whereas you want the first value to be A9. Position your cursor up in the formula line right in front of the letter B. Delete B and type in A, since your lower value is in cell A9. Your formula should now look like: =NORMDIST(A9,C9,D9,TRUE). Press **Enter**.

7) **Creating the remaining formulas:** We still need values for P (x > value) and P (Lower < x < Upper).

    a. **P (x > value):** Position your cursor in cell F5. Since the total area under the curve is 1, to find the P (x > value) we just need to create the formula: 1 – P (x < value). In cell F5, type in the formula: =1-E5 and then press **Enter.**

    b. **P(Lower < x < Upper):** Position your cursor in cell F10. Since we can find the value of the area bounded by the lower and upper boundaries by subtracting P (x < Upper) – P (x < Lower), in cell F10 type the formula: =f8-f9 and then press **Enter.**

    c. **Your sheet should look like the one shown:**

	A	B	C	D	E	F
1	Worksheet to Compute Probabilities for a Normal Distribution					
2						
3	Given One Value					
4	Enter Value	Enter Mean	Enter SD		P(x < Value)	P(x > Value)
5	174	172	29		0.527491466	0.472508534
6						
7	Between Two Values					
8	Enter Lower Value	Enter Upper Value	Enter Mean	Enter SD	P(x < Upper)	0.472508534
9	150	170	172	29	P(x < Lower)	0.224039746
10					P(Lower < x < Upper)	0.248468787
11						

FINDING VALUES FROM KNOWN AREAS

Refer back to the worksheet that we created in section 6-2 for "Finding the Score When Given a Probability". We set this worksheet up so that it could be used with any normal distribution. Simply enter appropriate information for the exercise involved into this worksheet. For a non-standard normal, our mean and standard deviation do not have to be 0 and 1 respectively.

TO PRACTICE THESE SKILLS

Once you have the tables above set up, you can quickly change the values that you input. Excel will automatically update the values in the columns containing the formulas. To ensure that you can always retrieve the original table that you set up by following the instructions above, it is always best to copy the table to another worksheet before you begin modifying it.

After you complete an exercise, you may want to copy and paste your completed table to another location in your worksheet. After you have copied the table, you should activate the cell where you want your table to begin. Then click on **Edit, Paste Special**, and click in the bubble by the word **Values**. Then click on **OK**. If you activate one of the cells where you had entered a formula originally, you will notice that now only the value shows up. The original formula is no longer active in the copied table. You can go back to the original table to compute the value for the next problem.

1) Try using the table you set up to find Probabilities using the NORMDIST function for exercises 5 through 8 and 13, 14, and 19 from section 6-3 Basic Skills and Concepts in your textbook.

2) Try using the table you set up to find Values from Known Areas, and modify the numbers as appropriate to address exercises 9 through 12 and exercises 17, 22, 23 and 24 from Section 6-3 Basic Skills and Concepts in your textbook.

3) Work with the appropriate table, and modify the numbers as appropriate to address exercises 15, 16, and 20 from Section 6-3 Basic Skills and Concepts in your textbook.

SECTION 6-4: SAMPLING DISTRIBUTIONS AND ESTIMATORS

This manual does not contain any new material specifically associated with section 6-4. You may find it helpful to review the material on finding the mean from a probability distribution presented back in section 5-2 of this manual.

SECTION 6-5: THE CENTRAL LIMIT THEOREM

In this section, we will use Excel to help us visualize the Central Limit Theorem. We can also modify the worksheet we created earlier to compute probabilities for a normal distribution so that we can compute probabilities for situations where the Central Limit Theorem needs to be used.

VISUALIZING THE CENTRAL LIMIT THEOREM

To help us see some of the major concepts behind the Central Limit Theorem, we will generate a table containing 1500 randomly generated digits 0, 1, 2, 3, …., 9. We will set the worksheet up in 30 columns and 50 rows.

1) **Name your worksheet:** In your worksheet, type "Randomly Generated Digits" in cell A1.

2) **Use the Random Number Generator:** Position your cursor in cell A2, and click on **Insert, Function.** In the "Search for a Function" box, type in random number and press **Enter.** You should see **RANDBETWEEN** listed in the "Select a Function" box. Make sure this is highlighted, and press **ENTER.** In the dialog box, enter 0 for the Bottom value and enter 9 for the Top value. Then click on **OK.**

3) **Use the fill handle to create our table of values:** We want to create 30 columns of randomly selected values, where each column has 50 rows. Make sure your cursor is positioned in cell A2, and using the fill handle, pull your mouse horizontally until you are out in column AD. Then release your mouse. Access the fill handle again, and pull down vertically until you are in row 51. You should now have a table of 1500 randomly generated digits.   Your table should look

different from others, in that the values in each cell are being randomly generated. Just be aware of this if you are comparing your table to another classmate's table.

4) **Find the mean of each row:** Position your cursor in cell AF1 and type in "ROW MEAN".

    a. **Access the AVERAGE function:** Position your cursor in cell AF2, and click on **Insert Function** (or click on the function icon in the toolbar). In the "Search for a function" box, type in Mean, and press **Enter**. You should see **AVERAGE** highlighted in the "Select a function" box. Click on **OK**. In the input box by Number 1, type in the range "A2:AD2", and click on **OK**. In cell AF2, you should now see the average of the thirty numbers in row 2.

    b. **Use the fill handle to copy the formula:** Position your cursor in cell AF2. Access the fill handle, and pull the mouse down to cell AF51. You should now have the average for each row of 30 values.

5) **Stabilize your table:** You may notice that as you work with your columns and cells, the table of numbers you generated changes. To stabilize the set of data you are working with, highlight the columns containing your values and your means and click on **Edit, Copy**. Move to a new worksheet and position your cursor in cell A1. From the toolbar, click on **Edit, Paste Special**, and then click on the bubble by **Values**. You will now have a stable table, since the cells are representing numbers now rather than formulas. Your pasted table may very well be quite different from the table that you chose to copy! From here out, the copied table will remain stable, as it is no longer associated with the random number generator.

6) **Create a histogram for the 1500 digits found in your table:** Use the Upper Class Limits of 0, 1, 2, 3, ..., 9 so that we get a distribution that shows how many of each digit occurred in the data. Refer back to Chapter 3 as necessary to review how to create a frequency distribution and histogram similar to the one shown below. Remember, since your values were randomly generated, your information does not have to be exactly the same as that shown below. You should see clearly that your histogram does not appear bell shaped.

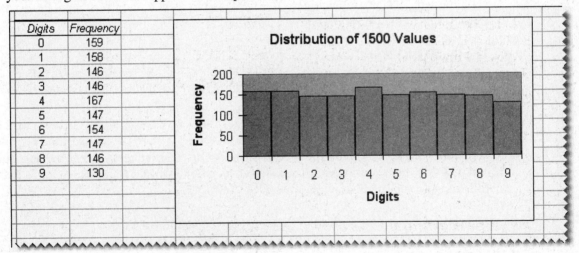

Digits	Frequency
0	159
1	158
2	146
3	146
4	167
5	147
6	154
7	147
8	146
9	130

Distribution of 1500 Values

7) **Create a histogram for the 50 Sample Means:** To make it easy to compare the pictures, use the "bins" or upper class limits from 0 to 9 in intervals of 0.5. Although the histogram of sample means may not look entirely bell shaped, it is definitely moving in that direction.

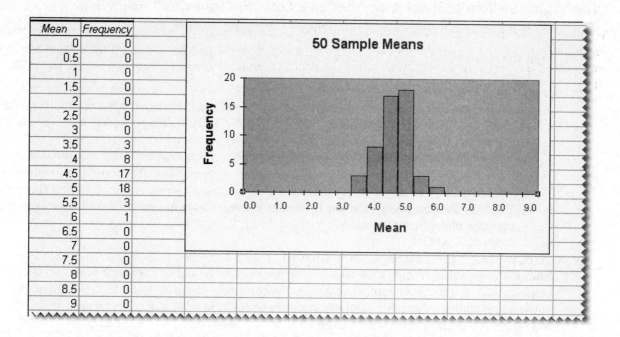

Mean	Frequency
0	0
0.5	0
1	0
1.5	0
2	0
2.5	0
3	0
3.5	3
4	8
4.5	17
5	18
5.5	3
6	1
6.5	0
7	0
7.5	0
8	0
8.5	0
9	0

8) **Computing the Means and Standard Deviations:** In order to see more of the results of the Central Limit Theorem, we want to compute the mean of both the original 1500 values as well as the mean of the 50 sample means. We will also want to compute the Standard Deviations for each set of numbers.

   a. **Use the AVERAGE Function:** Set up an area like the one shown below, and use the AVERAGE function for the entire 1500 values (A2:AD51), as well as for the 50 sample means (AF2:AF51).

   b. **Use the STDEVP Function:** Since we are considering the 1500 values a population, we should use the STDEVP function to find the standard deviation of this population. Apply the function to the values in cells (A2:AD51).

   c. **Use the STDEV Function:** Since we are considering a samples when we compute the sample means, we would use the STDEV function to find the standard deviation of these sample means. (AF2:AF51)

   d. **Computing the Theoretical Standard Deviation:** The Central Limit Theorem says that the standard deviation of the sample means should be the standard deviation of the population divided by the square root of the sample size. Set up a formula that shows the appropriate cell address where your population standard deviation is located, and which looks like the following: = (cell address of population SD)/sqrt(30).

9) **Notice the relationships:** Notice that the mean of your population and the mean of your sample means comes out to be the same (or very, very close), while the standard deviation of your sample means is fairly close to your computed $\sigma / \sqrt{n}$.

Mean of 1500 Values	4.388		Mean of 50 Sample Means	4.388
SD of 1500 Values	2.845134		SD of 50 Sample Means	0.506142
			Theoretical SD	0.519448

COMPUTING PROBABILITIES THAT INVOLVE THE CENTRAL LIMIT THEOREM

We have previously constructed worksheets that allow us to enter particular values and then use appropriate functions in Excel to compute the probabilities associated with these values. We can modify the worksheet we created for computing probabilities for normal distributions (See section 6-3) so that we can use it for situations involving sample means.

1) **Copy and paste your previous worksheet:** Assuming you have previously created and saved the worksheet called: "Worksheet to Compute Probabilities for a Normal Distribution" (see section 6-3), copy and paste that worksheet to another worksheet.

2) **Insert a column for the sample size:** In order to use the Central Limit Theorem, we need to be able to call on the sample size. If your worksheet is set up as the one from section 6-3, click at the top of column E. Then click on **Insert, Column.** You should see a column inserted in the worksheet to the left of your probability headings.

3) **Change your column headings:** You will need to make a number of changes to your headings. Change your headings to match the ones shown below. Notice that x changes to x bar to indicate you are talking about the probability of a sample mean. Also, since we don't have any particular values entered yet, the cells that are defined by a formula show: #NUM! Don't worry, this will change once we enter values in our shaded cells.

	A	B	C	D	E	F	G
1	Worksheet to Compute Probabilities for the CLT						
2							
3	Given One Value						
4	Enter sample value	Enter Population Mean	Enter Population SD	Enter Sample Size		P(x bar < Value)	P(x bar > Value)
5						#NUM!	#NUM!
6							
7	Between Two Values	⊕					
8	Enter Lower Value	Enter Upper Value	Enter Population Mean	Enter Population SD	Enter Sample Size	P(x bar < Upper)	#NUM!
9						P(x bar < Lower)	#NUM!
10						P(Lower < x bar < Upper)	#NUM!
11							

4) **Change your formulas:** There are 3 places where we need to adjust our previously entered formulas, cell F5, G8 and G9.

   a. **Cell F5:** We need to change the standard deviation that was originally used when we set the worksheet up for the normal probabilities. We want to use a standard deviation which would be the population standard deviation (cell C5) divided by the square root of the sample size (cell D5). We can put our cursor in cell F5 and modify the formula in the formula bar so that we see: =NORMDIST(A5,B5,C5/SQRT(D5),TRUE). Then press **Enter.**

   b. **Cell G8:** Again, we need to change the standard deviation. Your formula should look like: =NORMDIST(B9,C9,D9/sqrt(E9),TRUE)

   c. **Cell G9:** This needs a similar change to the standard deviation, and should look like: =NORMDIST(A9,C9,D9/SQRT(E9),TRUE)

5) **Input some specific values:** Let's use the example on Water Taxi Safety, part b. You are given a sample size of 20. You were told that the Mean of the population was 172 and the standard deviation of the population was 29.

  a. **Suppose we wanted P(xbar > 174):** Enter 174 for your sample value, 172 for your population mean, 29 for your population SD, and 20 as your sample size. You should see that the probability that you would get a sample mean greater than 174 is about .37888. This means that there is approximately a 37.88% chance that the mean weight of 20 randomly selected men would be greater than 174.

  b. **Suppose we wanted P (172 < xbar < 175):** In the second part of the worksheet, enter 172 as your lower value, 175 as your upper value, 172 as your population mean, 29 as your population SD, and 20 as your sample size. You should see that the probability that you would get a sample mean greater than 172 but less than 175 is about .1782. This means that there is approximately a 17.82% chance that the mean weight of 20 randomly selected men would be between 172 and 175.

	A	B	C	D	E	F	G
1	Worksheet to Compute Probabilities for the CLT						
2							
3	Given One Value						
4	Enter sample value	Enter Population Mean	Enter Population SD	Enter Sample Size		P(x bar < Value)	P(x bar > Value)
5	174	172	29	20		0.621119823	0.378880177
6							
7	Between Two Values						
8	Enter Lower Value	Enter Upper Value	Enter Population Mean	Enter Population SD	Enter Sample Size	P(x bar < Upper)	0.678186903
9	172	175	172	29	20	P(x bar < Lower)	0.5
10						P(Lower < x bar < Upper)	0.178186903
11							

TO PRACTICE THESE SKILLS

You can apply the skills you learned in this section by working on the following exercises.

1) Repeat the exercise presented in this section, but using a table with 100 rows of 30 digits. Compare the histograms you create from your new data set to those presented in this section.

2) Many of the exercises in Section 6-5 of your textbook deal with finding probabilities. Use the worksheets: "Worksheet to Compute Probabilities for a Normal Distribution" and "Worksheet to Compute Probabilities for the CLT" as appropriate to work on the exercises beginning with # 5 from section 6-5 Basic Skills and Concepts in your textbook

SECTION 6-6: NORMAL DISTRIBUTION AS APPROXIMATION TO BINOMIAL

If we have a binomial distribution where $np \geq 5$ and $nq \geq 5$, we can approximate binomial probability problems by using a normal distribution. The material below helps you see a clear demonstration that this approach will work. We will plot three different binomial probability distributions to see that as the sample size increases, our distribution appears to look more and more like a normal distribution.

In a new worksheet, we will create three binomial probability distributions:

- Let the first distribution have n = 10 and p = 0.5
- Let the second distribution have n = 25 and p = 0.5
- Let the third distribution have n = 50 and p = 0.5

1) **Refer back to section 5-3 to create Binomial Distributions**: Use the previous directions, and create your pairs of columns for the random variable and the associated probabilities in columns A& B; D & E; and G & H.

2) **Create the graphs of your 3 distributions:** You will be using the **Line-Column** option in the Chart Wizard. Click on the **Chart** icon, and click on the **Custom Types** tab. Scroll down until you see the **Line-Column** option. Click on this option, and then click on **Next.**

3) **For Graph 1:** The steps below take you through inputting appropriate information into the chart wizard.

   a. **Data Range:** Select the cells containing the probabilities in column B.

   b. **Filling out information in Series:** Click on the Series tab. We need to input the Category (X) axis labels. You want to use your x values here, so select the cells where you have your random variables (column A). We also have to add the second series. Underneath the **Series** box, click on **Add**. You should see another Series name in the box. Delete the information that is in the **Values** box, and select the probabilities

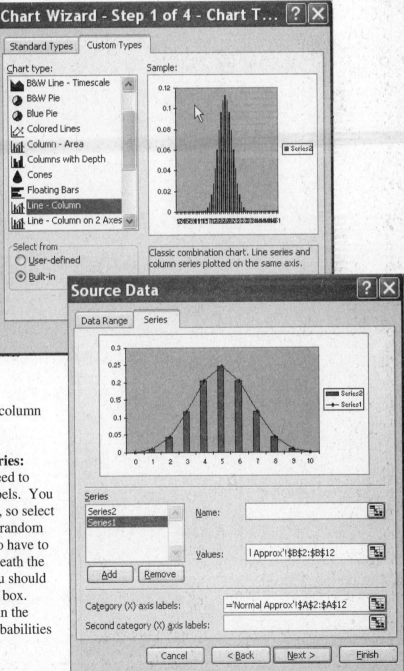

that are in column B. Your dialog box should look similar to the one shown.

c. **Enter the appropriate names:** Click on **Next**, and enter the appropriate information for the Chart Title and the axes names. Then click on **Next,** and choose whether you want your chart in a new sheet, or in the same sheet as your data.

d. **Modify the Chart:** Once your chart appears, double click in the **Plot Area**. Click on **Options** in the dialog box, and change your gap size to 0.

4) **Repeat this process to create the pictures for the other two binomial distributions**. Your pictures should appear as the three shown:

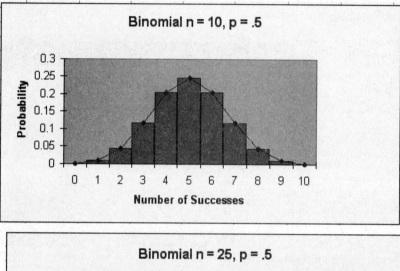

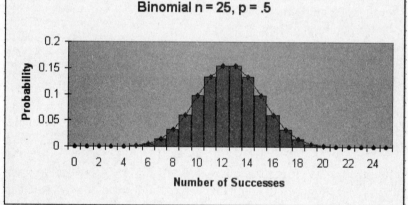

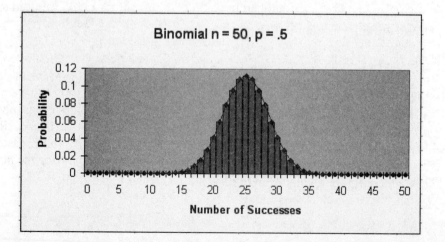

5) **Notice what happens to the graphs:** What you should notice about each of these successive pictures is that the areas in the bars, which represent the probabilities for each random variable, more closely approximates the area contained under the curve. If you think of the curve as representing a normal curve, than you can clearly see that the binomial probability distribution is more closely approaching a normal distribution.

6) **Compute the probabilities:** We can see that the probabilities we find using the normal distribution will closely match those that we can generate from the binomial distribution. To see this clearly, we will use the case of n = 50 and p = 0.5.

   a. **Add the binomial probabilities:** Using the values in our binomial distribution table, we can find P(x ≥ 30) by adding up the probabilities from x = 30 to x = 50. This produces a value of .101319.

   b. **Finding cumulative binomial probabilities without the entire table:** Most often, unless we were graphing the entire binomial probability distribution, we would not typically generate the entire list of values. Suppose we just wanted to quickly compute the probability above. In Excel, access the

BINOMDIST function, and fill out the dialog box as shown. Notice that Cumulative is marked as True. This tells Excel to add up the probabilities for x = 0 to x = 29. Since we want the probability that x is greater than or equal to 30, subtract this value from 1 to find the probability of .101319.

   c. **Find the probability using the normal distribution:** If we want to use a normal distribution, we will need to compute the mean and standard deviation using the formulas for a binomial distribution. You should find the value for the mean is 25 and the standard deviation is 3.535534. As outlined in section 6-6 of your textbook, we need to use a continuity correction when using the normal distribution to approximate the binomial. Since we want to find the probability of getting a value greater than or equal to 30, we should find P(x ≥ 29.5). Use the **NORMDIST** function on Excel, or use the worksheet you created previously. Remember that Excel returns a value that represents the area under the normal distribution curve to the left of the value. In order to find the probability we really want, we will need to subtract this value from 1. The **NORMDIST** function returns a value of 0.898454. Subtracting this value from 1 produces a probability of .101546. Recall that the probability using our binomial information is .101319. If we were to increase our sample size, we would find that we get even closer approximations when using the normal distribution to approximate the binomial.

TO PRACTICE THESE SKILLS

You can apply the skills you learned in this section by working on the following exercises.

1) Create the picture showing the Line-Column graph for a binomial probability distribution where n = 100 and p = 0.5. Compare your picture to those for n = 10, 25 and 50 found in this section.

2) Use the **NORMDIST** function on Excel, and the continuity correction to find the probabilities for the odd numbered exercises 5 – 12 in Section 6-6 Basic Skills and Concepts in your textbook. To recall how to work with the NORMDIST function, you may want to refer back to instructions found in Section 6-3 of this manual.

3) Use the **BINOMDIST** function on Excel, and, when appropriate, the **NORMDIST** function with the continuity correction to find the probabilities asked for in the odd numbered exercises 13 – 32 from Section 6-6 Basic Skills and Concepts in your textbook.

SECTION 6-7: DETERMINING NORMALITY

Oftentimes we want to know whether the data we are working with is normally distributed. We have already learned how to create histograms for sample data. From our histogram, we can reject normality if the histogram departs dramatically from bell shape.

An alternate way to determine normality is to construct a normal quantile plot (or normal probability plot) for the data. In a normal probability plot, the observations in the data set need to be ordered from smallest to largest. These values are then plotted against the expected z scores of the observations calculated under the assumption that the data are from a normal distribution. When the data are normally distributed, a linear trend will result. A nonlinear trend suggests that the data are non-normal.

We can use the **DDXL** Add-In to generate a normal probability plot. We will create a model using the data for **Diet Coke** found in Data Set 12: Weights and Volumes of Cola.

1) **Load the data from the CD that comes with your book (COLA.XLS), or from the website:** Copy the column showing the 36 weights of Diet Coke (CKREGWT) into column A of a new Excel worksheet.

2) **Access DDXL:** Click on **DDXL** on your menu bar.

3) **Click on Charts and Plots**. Click on the down arrow on the Function type box, and click on **Normal Probability Plot.** Click on the pencil icon for **Quantitative Variable**, and enter the range of values for your data. If your column title appears in cell A1, and your data started in cell A2, you range will be entered as "A2:A37".

4) Click on **OK.** You should see the following information on your screen.

Data Desk® 6.1 Viewer - Untitled

File Edit Data Special Help

$VAR1 Normal Prob. Plot

Normal Probability Plot Guidance

A normal probability plot is a good way to assess how close a variable's distribution is to a Normal distribution. If the diagonal stripe of points in the probability plot follows a straight line, then the data are normally distributed. If points at one edge or the other jump away from the overall trend, they may be outliers deserving special attention. If you have defined a label variable, select the Query tool from the Tools palette and click on any point to identify it.

The **More on Probability Plots** button provides more information on how probability plots are made and how to interpret them.

Summary statistics provide concise descriptions of the numbers in a variable and permit simple comparisons of the variable to other variables or to external standards. They can be a helpful complement to the normal probability plot. The **Summary Details** button defines the summary statistics displayed in the table.

More on Probability Plots

Summary Details

$VAR1 Summary

Count	36
Mean	0.785
Median	0.785
Std Dev	0.00439
Variance	1.928e-5
Range	0.0165
Min	0.776
Max	0.792
IQR	0.0057
25%	0.782
75%	0.788

TO PRACTICE THESE SKILLS

You can apply the skills from this section by working with exercises 9, 10 and 11 from section 6-7 Basic Skills and Concepts in your textbook. For each exercise, you should open the file in an Excel workbook from the CD that comes with your book or from the Addison Wesley website (www.aw.com/triola). The file names are listed below:

For exercise 9, BMI data is in the file named MHEALTH.XLS and FHEALTH.XLS.
For exercise 10, Weights of Pennies data is in the file named COINS.XLS.
For exercise 11, Precipitation data is in the file named WEATHER.XLS.

CHAPTER 7: ESTIMATE AND SAMPLE SIZES

SECTION 7-1: OVERVIEW 97

SECTION 7-2: ESTIMATING A POPULATION PROPORTION 97

DETERMINING A CONFIDENCE INTERVAL FOR A POPULATION PROPORTION: 97

TO PRACTICE THESE SKILLS 99

SECTION 7-3: ESTIMATING A POPULATION MEAN: σ KNOWN 99

DETERMINING CONFIDENCE INTERVALS 100

CONFIDENCE INTERVALS WITH DDXL 101

TO PRACTICE THESE SKILLS 102

SECTION 7-4: ESTIMATING A POPULATION MEAN: σ NOT KNOWN 102

DETERMINING CONFIDENCE INTERVAL USING A T DISTRIBUTION 103

CONFIDENCE INTERVALS WITH DDXL 104

TO PRACTICE THESE SKILLS 104

SECTION 7-1: OVERVIEW

It is not unusual for us to be in a situation where we do not know, or have access to, the parameters that are used to describe a population. In these instances we need to resort to using information contained in a given sample. If we can identify a numerical value that describes a given sample, then this value can also be used to estimate the corresponding descriptor for the population. **Confidence intervals** are important in statistics because they allow you to gauge how accurately a sample parameter approximates the same parameter with respect to the population. The confidence interval consists of a range (or interval) of values instead of just a single value. The confidence interval also contains a probability. This probability value tells you the likelihood that you have an interval that actually contains the value of the unknown population parameter. Components of a confidence interval include a lower limit and an upper limit for the parameter under consideration, as well as a probability value.

Excel does not have a built in function that automatically calculates the confidence interval. Instead, we will need to rely on some of the functions we have already used in Excel to help us build a confidence interval. In addition to creating a confidence interval with Excel we will make use of the add-in DDXL. This add-in was supplied on the CD that came with your textbook. If you did not load the DDXL add-in or cannot find it on the computer you are working with please go back to Chapter 2, Section 1 and follow the instructions for loading DDXL.

The following list contains an overview of the functions we will be utilizing in this chapter.

CONFIDENCE: Returns the confidence interval for a population mean.
CONFIDENCE (alpha, standard_dev, size) where alpha refers to the significance level used to compute the confidence interval, standard_dev is the population standard deviation for the data range and size is the sample size.

TINV: Returns the inverse of the Student's t-distribution for the specified degrees of freedom.
TINV (probability, deg_freedom) where probability is the probability associated with a two tailed Student's t distribution and deg_freedom is a positive integer indicating the number of degrees of freedom needed to characterize the distribution.

SECTION 7-2: ESTIMATING A POPULATION PROPORTION

As stated in the Overview, Excel does not produce confidence interval estimates for proportions. It will be necessary for us to use the **DDXL** add-in that came as a supplement to your textbook to do this. **If you have not loaded DDXL yet please do so before going any further.** The instructions for adding this feature to Excel can be found in Chapter 2, Section 1 of this manual. If you are unsure as to whether or not you have already added DDXL check the menu bar in Excel to see if it featured on the menu bar as seen below.

DETERMINING A CONFIDENCE INTERVAL FOR A POPULATION PROPORTION:

To determine the confidence interval for a population proportion using **DDXL** we will begin by considering the **"Touch Therapy"** results found in the Chapter Problem at the start of Chapter 7.

1) Label cells A1 and A2 as shown in the Excel worksheet on the next page. Enter the number of trials in cell B1 and the number of successes in cell B2.

2) Click on **DDXL** and highlight **Confidence Intervals** and click.

3) Select **Summ 1 Var Prop Interval** from the Confidence Intervals Dialog box that opens.

4) Click on the **pencil icon** under the heading **"num successes"** and enter the cell address B2. **Click on** the **pencil icon** under the heading **"num trials"** and enter the cell address B1.

Screen shots for steps (1) through (4) can be seen as follows:

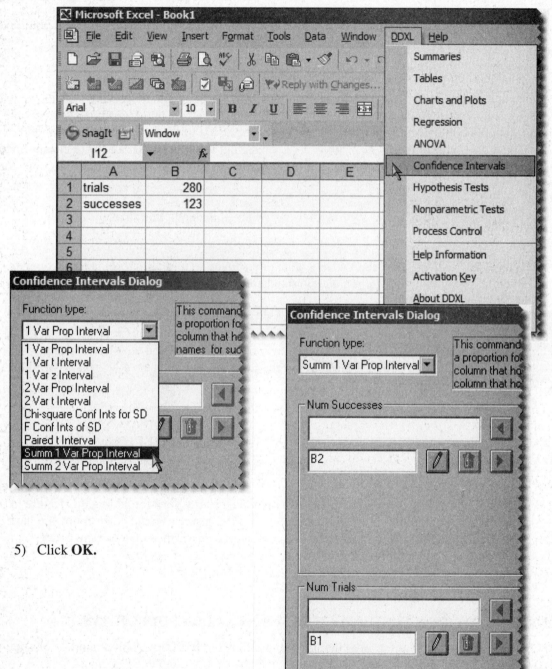

5) Click **OK.**

In the Proportion Interval Setup dialog box

a. Select the appropriate level of
confidence, in this case **95%**.

b. Click on **Compute Interval**.

A summary dialog box
similar to the one shown
here displays the
confidence interval for
the population
proportion.

It is possible to **copy and paste your DDXL results** to your Excel worksheet.

1) Click on the title bar of the window you wish to copy into Excel.

2) From the **Edit** menu choose **Copy Window**.

3) Switch to Excel and choose **Paste** from the **Edit** menu or use the **Paste Function** feature.

TO PRACTICE THESE SKILLS

You can practice the technology skills learned in this section by working through the following problems
found in your textbook.

1) Use **DDXL** to work through problems 21 and 23 in Section 7-2 Basic Skills and Concepts. Note that the
problem asks for the confidence interval as a percentage. DDXL will return a decimal value. You will
need to rewrite your results in the appropriate format.

2) Use **DDXL** to work through problem 31 in Section 7-2 Basic Skills and Concepts. Note that you will
have to determine the number of successes before you begin.

3) Use **DDXL** to work through problem 35 in Section 7-2 Basic Skills and Concepts.

4) Use the data found in **Data Set 13** in appendix B of your textbook or the data file "M&M" found on the
CD data disk to work through problem 45 in Section 7-2 Basic Skills and Concepts.

SECTION 7-3: ESTIMATING A POPULATION MEAN: σ KNOWN

In Section 7- 2 we used a sample proportion as the best point estimate of the population proportion. In this
section we will apply some of the same ideas to $\overline{X}$ as the best point of the population mean μ. A confidence
interval gives us information that enables us to better understand the accuracy of the point estimate of a

population parameter. The formula for the confidence interval for the population mean when the standard deviation σ is known is given by $\overline{X} - E < \mu < \overline{X} + E$ where $E = z_{\alpha/2} \cdot \dfrac{\sigma}{\sqrt{n}}$

To determine a confidence interval for the mean in Excel you need to know the value for the sample mean, $\overline{X}$, and for the standard deviation, σ. It is relatively easy to find both of these values using the built in statistical functions in Excel or from Descriptive Statistics found in the Data Analysis Tools.

DETERMINING CONFIDENCE INTERVALS

As we work through the process of building a confidence interval we will use the data found in **Data Set 18** in Appendix B (Homes Sold In Dutchess County) of your textbook or the data file "HOMES.XLS" found on the CD data disk.

- **Copy the column** titled **Selling Price** (dollars) into a new worksheet.

- **Use the Insert function** feature of Excel to determine the sample mean and standard deviation.

To find the **confidence interval** we make use of the **Insert Function** feature introduced in previous chapters.

1) In the **Insert Function** dialog box select the **Statistical** function.

2) Scroll through the list of Function names until you see **CONFIDENCE**. Click on this function name.

3) Click **OK**.

4) This opens a Confidence dialogue box. Assuming a degree of confidence of 95% we can fill in the information required.

  a. With a degree of confidence of 95%, α = **0.05.**

  b. Enter the standard deviation and the size of the sample.

5) Your dialog box should look similar to the screenshot found on the next page.

The value returned is called the **margin of error** (or **maximum error**) and is denoted by the letter E in the formula presented at the beginning of this section. You now have all of the information you need to create a confidence interval.

1) Determine the upper confidence interval limit ($\overline{X} + E$) and the lower confidence interval limit ($\overline{X} - E$).

2) Using the general format $\overline{X} - E < \mu < \overline{X} + E$ used to display a confidence interval substitute in the values found in step #1 above.

C	D
mean	342672.5
standard deviation	83236.66
margin of error	25794.82
upper class limit	368467.3
lower class limit	316877.7

   a. The lower limit is found using the formula = (D1-D4)

   b. The upper limit is found by using the formula = (D1 + D4)

3) This will give you a confidence interval $316877.7 < \mu < 368467.3$.

CONFIDENCE INTERVALS WITH DDXL

You may also determine a confidence interval for a population mean when σ is known by using the DDXL add-in. You are strongly encouraged to try both methods (Excel functions and the DDXL add-in) in order to determine the advantages and disadvantages of each method.

To use DDXL to determine a confidence interval it will be necessary for you to enter your data in a single column. Earlier in the chapter we used the data found in **Data Set 18** in Appendix B (Homes Sold In Dutchess County) of your textbook or the data file "HOMES.XLS" found on the CD data disk. Open this data file.

1) **Copy the column** titled **Selling Price** (dollars) into a new worksheet.

2) **Use the Insert function** feature of Excel to determine the sample mean and standard deviation.

3) Click on **DDXL.**

4) Highlight **Confidence Intervals** and click.

5) Select **1 Var z Interval**.

6) Click on the **pencil icon** and enter the range of data (for the HOMES.XLS data this range will be A2:A41).

7) Click **OK.**

8) In the dialog box that looks like the one seen on the right

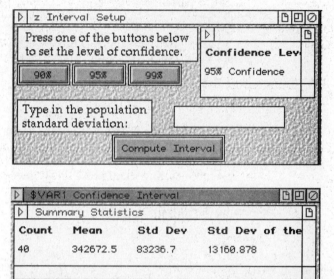

    a. Select the appropriate level of confidence, in this case **95%**.

    b. Enter the standard deviation for the sample.

    c. Click on **Compute Interval**.

A summary dialog box similar to the one shown on the right displays the confidence interval for the population proportion. Compare this to the result we determined previously in Excel by finding the margin of error (316877.7 < μ < 368467.3).

As with our previous DDXL problem, it is possible to copy and paste your results to your Excel worksheet.

TO PRACTICE THESE SKILLS

The problems following Section 7-4 in this manual will help you to practice the technology skills introduced in Sections 7-3 & 7-4 of this chapter. Use both Excel and DDXL when working through these problems in order to become familiar with both methods.

SECTION 7-4: ESTIMATING A POPULATION MEAN: σ NOT KNOWN

In this section we will present a method for determining a confidence interval estimate for the population mean, μ without the requirement that σ be known. The student t distribution is used to estimate the population mean. The sample confidence interval is $\overline{X} - E < \mu < \overline{X} + E$ where $E = t_{\alpha/2} \cdot \dfrac{s}{\sqrt{n}}$.

Note that the sample standard deviation s, replaces the population standard deviation σ in the formula. To determine the small sample confidence intervals or the population mean with Excel, use the **TINV** function to determine the appropriate t **values**.

In this section we will work with the data found in Data Set 1 in Appendix B (Health Exam Results), file name MHEALTH.XLS. Use the sample pulse rates found in this data set.

1) **Copy this column (Pulse)** to a blank worksheet.

2) **Determine** the **sample mean** and **standard deviation**.

We will use this information to determine

    a.   The margin of error E.

    b.   The confidence interval for μ.

These can be determined easily using the built in functions in Excel.

DETERMINING CONFIDENCE INTERVAL USING A T DISTRIBUTION

To determine the *t* **value** we will follow the same method used to find the confidence interval outline in Section 7–3.

To find the TINV confidence interval we need to use the **Insert Function** feature.

1)   In the **Insert Function** dialog box select the **Statistical** function.

2)   Scroll through the list of Function names until you see **TINV**. Click on this function name.

3)   Click **OK.**

The following dialogue box will open in your Excel workbook.

Function Arguments

TINV

Probability   = number

Deg_freedom   = number

=

Returns the inverse of the Student's t-distribution.

Deg_freedom is a positive integer indicating the number of degrees of freedom to characterize the distribution.

Assuming a degree of confidence of 95% we can fill in the information required. Recall that with a degree of confidence of 95%, the probability = 0.05, the deg_freedom is 40.

mean	69.400
st dev	11.297
sample size	41
t distribution	2.021
error (E)	3.565891
upper limit	72.966
lower limit	65.834

4)   This returns a **t-distribution** of 2.021.

5)   **To find our confidence interval** complete the information as seen on the Excel worksheet on the right.

    a. **To determine the value of E** use the formula $E = t_{\alpha/2} \cdot \dfrac{s}{\sqrt{n}}$.

    b. Use $\overline{X} - E < \mu < \overline{X} + E$ to find the upper and lower limit of the confidence interval.

CONFIDENCE INTERVALS WITH DDXL

You may also determine a confidence interval for a population mean when σ is not known using DDXL. You are strongly encouraged to try both methods (Excel functions and the DDXL add-in) in order to determine the advantages and disadvantages of each method.

To use DDXL to determine a confidence interval when σ is not known simply follow the instructions outlined in the previous section with one change – choose the **1 Var t Interval.**
The following result will be returned:

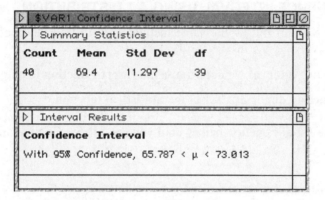

TO PRACTICE THESE SKILLS

The following problems will help you to practice the technology skills introduced in Sections 7-3 & 7-4 of this chapter. Use both Excel and DDXL when working through these problems in order to become familiar with both.

1) Use the data found in Data Set 2 in Appendix B of your textbook or the data file "BODYTEMP" found on the CD data disk that accompanies your book to work through problem 18 in Section 7-4 Basic Skills and Concepts.

2) Use Excel and DDXL to work through problem 24 in Section 7-4 Basic Skills and Concepts.

3) Use Excel and DDXL to work through problem 28 in Section 7-4 Basic Skills and Concepts.

CHAPTER 8: HYPOTHESIS TESTING

SECTION 8-1: OVERVIEW **106**

SECTION 8-2: TESTING A CLAIM ABOUT A PROPORTION **106**

Z TEST FOR ONE VARIABLE PROPORTION TEST: **107**

TO PRACTICE THESE SKILLS **108**

SECTION 8- 3: TESTING A CLAIM ABOUT A MEAN: σ KNOWN **108**

TESTING CLAIMS ABOUT A POPULATION MEAN μ (σ KNOWN) **108**

TO PRACTICE THESE SKILLS **109**

SECTION 8- 4: TESTING A CLAIM ABOUT A MEAN: σ NOT KNOWN **109**

TESTING CLAIMS ABOUT A POPULATION MEAN μ (σ NOT KNOWN) **109**

TO PRACTICE THESE SKILLS **110**

SECTION 8-1: OVERVIEW

This chapter will look at the statistical process used for testing a claim made about a population. Chapter 7 used sample statistics to estimate population parameters; in this chapter we will use sample statistics to test hypotheses made about population parameters. While Excel has many built in statistical analysis tools available, it does not have a tool for hypothesis tests for the mean. We will use the DDXL add-in for the various hypotheses tests we will perform in this chapter.

SECTION 8-2: TESTING A CLAIM ABOUT A PROPORTION

Once you understand the different individual components of a hypothesis test (these components are explained in Section 8–2 of your textbook) you will be ready to use those components to test claims made about population proportions. We will utilize the DDXL add in to perform a z test of the hypothesis for a proportion.

Z TEST FOR ONE VARIABLE PROPORTION TEST:

In the Chapter Problem found at the beginning of Chapter 8 focused on the best way to go about finding a job it was noted that "Among 703 randomly selected workers, 61% got their jobs through networking." We will test the claim that "most (more than 50%) workers get their jobs through networking. This problem is outlined in the **Finding a Job Through Networking** example found in Section 8-3 of your textbook.

1) Begin by entering the number of workers surveyed and the population proportion that felt they got their jobs through networking into Excel:
   ($n = 703 \quad \hat{p} = 0.61$)

2) In order to work through this problem with DDXL you will need to enter the number of trials (n) and the number of successes ($n \cdot \hat{p}$). Compare your worksheet with the one shown on the right.

	A	B
1	n	703
2	proportion	0.61
3		
4	trials	703
5	successes	429

3) Select **DDXL** from the tool bar, scroll down to **Hypothesis Tests** and click.

4) From the **Hypothesis Tests Dialog** box select **Summ 1 Var Prop Test** from Function type.

5) Click on the pencil icon for **Num Successes** and enter the cell address for the number of workers who have successfully found a job. **Be sure to enter the cell address rather than the actual number of successes.**

6) Click on the pencil icon for **Num Trials** and enter the cell address of total number of workers surveyed. **Be sure to enter the cell address and not the actual number of trials.**

7) Click **OK**.

8) This will open a dialog box that will require additional information. Complete each of the four steps in the dialog box as outlined on the following page.

```
┌─┬──────────────────────────────────────────────┬──┐
│▷│  Proportion  Test  Setup                     │▫│
├─┴──────────────────────────────────────────────┴──┤
│  ┌──────────────────────────────────────────────┐ │
│  │Step 1: Enter the hypothesized population      │ │
│  │         proportion p0.                        │ │
│        ┌────────────────────────────┐             │
│        │          Set p0            │             │
│        └────────────────────────────┘             │
│  ┌──────────────────────────────────────────────┐ │
│  │Step 2: Set the significance (alpha) level.    │ │
│  ┌──────┐ ┌──────┐ ┌──────┐ ┌──────────┐         │
│  │ .01  │ │ .05  │ │ .10  │ │  Other...│         │
│  └──────┘ └──────┘ └──────┘ └──────────┘         │
│  ┌──────────────────────────────────────────────┐ │
│  │Step 3: Select an alternative hypothesis (Ha). │ │
│  ┌─────────────┐ ┌─────────────┐ ┌─────────────┐  │
│  │   p < p0    │ │   p ≠ p0    │ │   p > p0    │  │
│  └─────────────┘ └─────────────┘ └─────────────┘  │
│  ┌─┬────────────────────────────────────────┬──┐  │
│  │▷│ Settings                               │▫│  │
│  │   p0:                          0.5           │  │
│  │ Alpha:                         0.05          │  │
│  │   Ho:                    p = 0.5             │  │
│  │   Ha:    Select a 1- or 2-tailed test.       │  │
│  └──────────────────────────────────────────────┘  │
│  ┌──────────────────┐                              │
│  │Step 4: Compute.  │                              │
│        ┌────────────────────────────┐             │
│        │         Compute            │             │
│        └────────────────────────────┘             │
└────────────────────────────────────────────────────┘
```

a. **Step 1: Click on "Set p0".** We will enter the hypothesized test proportion found within the example ($p = 0.5$). Enter the value in the appropriate spot and click **OK.**

b. **Step 2:** Set the **significance level** by clicking on the appropriate value. In this case use a significance level of 0.05.

c. **Step 3:** Select the alternative hypothesis. In this problem the null hypothesis states H_0: $p = 0.5$. Therefore the alternative hypothesis is H_1: $p > 0.5$. Choose p > p0.

d. **Step 4:** Click on **Compute.**

9) **DDXL** presents the following results that include the test statistic, the P – value and a conclusion to reject the null hypothesis.

```
┌──────────────────────────────────────────────────────┐
│▷ $VAR1 Proportion Test                          ▫▫⊘  │
├──────────────────────┬────────────────────────────────┤
│▷ Summary Statistics  │▷  Test Summary                 │
│      n      703       │        p0:              0.5    │
│   p-hat    0.609      │        Ho:         p = 0.5     │
│  Std Dev   0.0189     │        Ha: Upper tail: p > 0.5 │
│                       │  z Statistic:           5.77   │
│                       │     p-value:          < .0001  │
│                       ├────────────────────────────────┤
│                       │▷  Test Results                 │
│                       │ Conclusion                     │
│                       │ Reject Ho at alpha =  0.05     │
├──────────────────────┴────────────────────────────────┤
│           ┌────────────────────┐                       │
│           │     New Test       │                       │
│           └────────────────────┘                       │
└──────────────────────────────────────────────────────┘
```

TO PRACTICE THESE SKILLS

You can apply the technology skills learned in this section by working on the following problems.

1) To practice testing claims about proportions work through problems 11, 13, 19 and 21 found in the Basic Skills and Concepts for Section 8-3 of your textbook.

2) Use Data Set 13 to work through problem 25 in the Basic Skills and Concepts for Section 8-3 of your textbook. This problem gives you an opportunity to start with actual data to test a claim about proportions.

3) Use Data Set 8 to work through problem 27 in the Basic Skills and Concepts for Section 8-3 of your textbook. This problem gives you an opportunity to start with actual data to test a claim about proportions.

SECTION 8-3: TESTING A CLAIM ABOUT A MEAN: σ KNOWN

In this section we will test claims made about a population mean μ, and we assume that the population standard deviation σ is known. We will work with Data Set 13 in Appendix B (M&M.XLS). Copy the column titled Red into a new worksheet and determine the sample mean and standard deviation.

TESTING CLAIMS ABOUT A POPULATION MEAN μ (σ KNOWN)

This problem is outlined in the *P*-Value Method example found in Section 8-4 of your textbook. We will test the claim of a production manager that the M&Ms have a mean that is actually greater than 0.8535 g.

1) Select **DDXL** from the tool bar, scroll down to **Hypothesis Tests** and click.

2) From the **Hypothesis Tests Dialog** box select **I Var z Test** from Function type.

3) Click on the pencil icon and list the range of cells that include your data. (In this example your range of cells should be A2:A14).

4) Click **OK**.

5) A dialog box will open. Complete each of the four steps listed in that dialog box as follows:

> **Step 1:** Click on "Set μθ and sd" . The hypothesized mean is 0.8635. The standard deviation is 0.0565. Enter these values in the appropriate spot and click **OK.**

> **Step 2:** Set the **significance level** by clicking on the appropriate value. In this case use a significance level of 0.05.

> **Step 3:** Select the **alternative hypothesis**. In this problem the null hypothesis states that the mean weight is 0.8635. The alternative hypothesis states that the mean weight is greater than 0.8635. Choose μ > μθ.

> **Step 4:** Click on **Compute**.

6) DDXL presents the following results that include the test statistic, the P – value and a conclusion to fail to reject the null hypothesis.

```
┌─────────────────────────────────────────────────────────────────────────┐
│ ▷ $VAR1 z Test                                                    ▣▣⊘ │
│ ┌────────────────────────────────┐┌─────────────────────────────────────┐│
│ │ ▷  Summary Statistics        ▣ ││ ▷  Test  Summary                 ▣ ││
│ │                                │││                                     ││
│ │       Count    13              │││         Ho:             μ = 0.864   ││
│ │        Mean    0.864           │││         Ha:  Upper tail: μ > 0.864  ││
│ │ Pop  StDev:    0.0565          │││ z Statistic:              0.002     ││
│ │                                │││    p-value:               0.499     ││
│ │                                │││                                     ││
│ │                                │└─────────────────────────────────────┘│
│ │                                │┌─────────────────────────────────────┐│
│ │                                ││ ▷  Test  Results                 ▣ ││
│ │                                ││ Conclusion                          ││
│ │                                ││ Fail to reject Ho at alpha = 0.05   ││
│ │                                ││                                     ││
│ └────────────────────────────────┘└─────────────────────────────────────┘│
│                        ┌──────────────────────┐                          │
│                        │      New Test        │                          │
│                        └──────────────────────┘                          │
└─────────────────────────────────────────────────────────────────────────┘
```

TO PRACTICE THESE SKILLS

You can practice the skills learned in this section by working through the following problems.

1) Use Data Set 13 found in Appendix B or in the data file M&M.xls to work through problem 9 in the Basic Skills and Concepts for Section 8-4 of your textbook.

2) Use Data Set 14 found in Appendix B or in the data file COINS.xls to work through problem 19 in the Basic Skills and Concepts for Section 8-4 of your textbook.

SECTION 8- 4: TESTING A CLAIM ABOUT A MEAN: σ NOT KNOWN

In this section we turn our attention to testing claims made about a population mean μ when the population standard deviation σ is not known. We will use the Student t distribution rather than the normal distribution used in the preceding section where σ was known. We will rely on the DDXL add-in to perform a t test of the hypothesis of the mean. This test compares the observed t test statistic to the point of the t distribution that corresponds to the test's chosen α level.

TESTING CLAIMS ABOUT A POPULATION MEAN μ (WITH σ NOT KNOWN):

Begin by entering the data found in the **Quality Control of M&Ms** example in Section 8-5. We will use this data and a significance level of α = 0.05 to test the claim that test the claim of a production manager that the M&Ms have a mean that is actually greater than 0.8535 g.

1) Select **DDXL** from the tool bar, scroll down to **Hypothesis Tests** and click.

2) From the **Hypothesis Tests Dialog** box select **I Var t Test** from Function type.

3) Click on the pencil icon and list the range of cells that include your data.

4) Click **OK**.

5) Complete each of the four steps listed in the dialog box as outlined below.

> **Step 1:** Click on **"Set $\mu\theta$ "**. Enter the values for the hypothesized population mean (H_o: $\mu = 0.8635$) and click **OK**.

> **Step 2:** Set the **significance level** by clicking on the appropriate value. In this case use a significance level of $\alpha = 0.05$.

> **Step 3:** Select the **alternative hypothesis**. In this problem the alternative hypothesis states test the claim of a production manager that the M&Ms have a mean that is actually **greater** than 0.8535 g. Therefore we choose $\mu > \mu\theta$.

> **Step 4:** Click on **Compute**.

6) **DDXL** presents the following results that include the test statistic, the P – value and a conclusion to fail to reject the null hypothesis.

TO PRACTICE THESE SKILLS

You can practice the technology skills learned in this section by working on the following problems.

1) Work through problem 17 in the Basic Skills and Concepts for Section 8-5 of your textbook.

2) Use Data Set 14 found in Appendix B or in the data file COINS.xls to work through problem 29 in the Basic Skills and Concepts for Section 8-5 of your textbook.

3) Use Data Set 1 found in Appendix B or in the data file MHEALTH.xls to work through problem 31 in the Basic Skills and Concepts for Section 8-5 of your textbook.

CHAPTER 9: INFERENCES FROM TWO SAMPLES

SECTION 9-1: OVERVIEW 112

SECTION 9-2: INFERENCES ABOUT TWO PROPORTIONS 112

DDXL – SUMM 2 VAR PROP TEST 113

DDXL - CONFIDENCE INTERVALS FOR PROPORTION PAIRS 113

TO PRACTICE THESE SKILLS 114

SECTION 9-3: INFERENCES ABOUT TWO MEANS: INDEPENDENT SAMPLES 114

t-TEST: TWO-SAMPLE ASSUMING UNEQUAL VARIANCES: 115

DDXL – 2 VAR t TEST 116

CONFIDENCE INTERVAL ESTIMATES 116

TO PRACTICE THESE SKILLS 117

SECTION 9-4: INFERENCES FROM MATCHED PAIRS 118

t-TEST: PAIRED TWO SAMPLES FOR MEANS 118

DDXL – PAIRED T TEST 119

DDXL – CONFIDENCE INTERVALS FOR MATCHED PAIRS 120

TO PRACTICE THESE SKILLS 121

SECTION 9-5: COMPARING VARIATION IN TWO SAMPLES 121

F-TEST: TWO SAMPLE FOR VARIANCES 121

TO PRACTICE THESE SKILLS 122

SECTION 9-1: OVERVIEW

In Chapters 7 and 8 we used sample data to construct confidence interval estimates of population parameters and to test hypotheses about certain population parameters. In each case we used one sample to form an inference about one population. In this chapter we will turn our attention to confidence intervals and hypothesis tests for comparing two sets of sample data. In this chapter we will utilize the capabilities if Excel as well as the DDXL add-in to test the hypothesis made about two population means.

- the Data Analysis **z-Test: Two Samples for Means.**

- the DDXL add-in Hypothesis Tests: **2 Var t Test**

t-Test: Paired Samples for Means
This analysis tool and its formula perform a paired two-sample student's t-test to determine whether a sample's means are distinct. This t-test form does not assume that the variances of both populations are equal. You can use a paired test when there is a natural pairing of observations in the samples, such as when a sample group is tested twice — before and after an experiment.

F Test Two Samples for Variances
This analysis tool performs a two-sample F-test to compare two population variances and to determine whether the two population variances are equal. It returns the p value of the one tailed F statistic, based on the hypothesis that array 1 and array 2 have the same variance.

t-Test: Two Samples Assuming Equal Variances
This test calculates a two sample Student t Test. The test assumes that the variance in each of the two groups is equal. The output includes both one tailed and two tailed critical values.

t-Test: Two Samples Assuming Unequal Variances.
This test calculates a two sample Student t Test. The test allows the variances in the two groups to be unequal. The output includes both one tailed and two tailed critical values.

SECTION 9-2: INFERENCES ABOUT TWO PROPORTIONS

When using sample data to compare two population proportions we will use DDXL and the **Summ 2 Var Prop Test.**   Since we are comparing two population proportions we will need to enter the number of successes as well as the number of trials for both Sample 1 and Sample 2 into Excel.

Using the information presented in the Chapter Problem as well as the **Is Surgery Better than Splinting?** Example found in Section 9-2 of your textbook into Excel:

	A	B	C
1		successes	trials
2	surgery	67	73
3	splinting	60	83

Use the following steps and a 0.05 significance level to test the claim that the success rate with surgery is better than the success rate with splinting.

DDXL – SUMM 2 VAR PROP TEST

1) Click on **DDXL** and choose **Hypotheses Tests** and **Summ 2 Var Prop Test**.

2) In the Hypothesis test dialog box click on the pencil icons and enter the appropriate cell address for the number of success as well as the address for the number of trials for surgery. Repeat this process to enter the cell addresses for the number of successes and the number of trials for splinting.

3) Click **OK**.

4) Follow these steps in the **Proportion Test Setup** dialog box:

   a. **Step 1:** This step is optional so you can skip it or set the difference at 0 (since we will only be testing claims that p1 = p2).

   b. **Step 2**: For this example set the significance level at 0.05.

   c. **Step 3**: Select the appropriate alternative hypothesis, in this case **p1 – p2 > p**.

   d. **Step 4**: Click on **Compute**.

5) DDXL returns the following results. Compare these results with those found in your textbook. DDXL returns the value for test statistic, the p –value as well as the conclusion to reject the null hypothesis.

DDXL - CONFIDENCE INTERVALS FOR PROPORTION PAIRS

We can construct a confidence interval estimate of the difference between population proportions for the sample date presented in the **Is Surgery Better than Splinting?** example using DDXL.

1) Click on **DDXL**, select **Confidence Interval – Summ 2 Var Prop Interval.**

2) In the dialog box click on each of the pencil icons and enter the appropriate cell addresses for the number of successes and trials for surgery. Repeat this process for the number of successes and trials for splinting.

3) Click **OK**.

4) A dialog box will open in order for you to set the appropriate confidence level. In this case we choose 90%.

5) Click on **Compute Interval.**

6) The following results are displayed.

```
┌─────────────────────────────────────────────────────────────────────┐
│ ▷ │ Confidence Interval for the Difference Between  and ç!    🗅🗖⊘   │
│ ▷ │ Summary Statistics       🗅                                       │
│         n1     73                                                     │
│      p-hat1    0.918                                                  │
│         n2     83              ┌──────────────────────────────────┐  │
│      p-hat2    0.723          ▷│ Interval Results            🗅   │  │
│    Difference  0.195           ├──────────────────────────────────┤  │
│      Std Err   0.0587          │ Confidence Interval              │  │
│         z*     1.64            │ With 90% Confidence, 0.0983 < p < 0.291 │
│                                └──────────────────────────────────┘  │
└─────────────────────────────────────────────────────────────────────┘
```

TO PRACTICE THESE SKILLS

You can practice the technology skills covered in this section by working on the following problems found in your textbook.

1) Work on problems 9, 15, and 17 in the Basic Skills and Concepts for Section 9-2 of your textbook. These problems focus on hypothesis testing and clearly identify the number of successes and trials for each sample.

2) Work on problems 23, 24 and 27 in the Basic Skills and Concepts for Section 9-2 of your textbook. These problems focus on determining confidence interval estimates and require you to determine the number of successes for each sample.

3) Use Data Set 10 found in Appendix B or in the data file BOSTRAIN.xls to work through problem 29 in the Basic Skills and Concepts for Section 9-2 of your textbook.

SECTION 9-3: INFERENCES ABOUT TWO MEANS: INDEPENDENT SAMPLES

In this section we will work with sample data from two independent samples to test the hypothesis made about two population means. We will also construct confidence interval estimates of the differences between two population means. We will outline the following methods:

- the Data Analysis **t-Test: Two-Sample Assuming Unequal Variances**

- the DDXL add-in Hypothesis Tests: **2 Var t Test**

Enter the data found in Section 9-3 in the **Discrimination Based on Age** example. Create a column of data containing the ages of unsuccessful applicants and a column of data containing the ages of successful applicants.

T-TEST: TWO-SAMPLE ASSUMING UNEQUAL VARIANCES:

1) Click on **Tools,** highlight **Data Analysis** and click.

2) From the Analysis Tools list box in the Data Analysis dialog box select **t-Test: Two-Sample Assuming Unequal Variances**.

3) Click on **OK.**

4) In the **t-Test: Two-Sample Assuming Unequal Variances** dialog box enter the following information:

   a. The range of cells containing the age of unsuccessful applicants in the **Variable 1 Range** box.

   b. The range of cells containing the age of successful applicants in the **Variable 2 Range** box.

   c. Enter 0 in the **Hypothesized Mean Difference** box or just leave it blank.

   d. Enter 0.05 in the **Alpha** box. This value is supplied in the problem.

   e. Determine where you wish to display the output.

   f. Click **OK**.

t-Test: Two-Sample Assuming Unequal Variances	?|X
Input	
Variable 1 Range:	$A$3:$A$25
Variable 2 Range:	$B$3:$B$32
Hypothesized Mean Difference:	
☐ Labels	
Alpha: 0.05	
Output options	
⦿ Output Range:	d1
○ New Worksheet Ply:	
○ New Workbook	
	OK
	Cancel
	Help

5) The following summary of information containing calculations for the t-test is added to your current Excel worksheet.

t-Test: Two-Sample Assuming Unequal Variances		
	Variable 1	Variable 2
Mean	46.95652174	43.93333333
Variance	52.13438735	34.61609195
Observations	23	30
Hypothesized Mean Difference	0	
df	42	
t Stat	1.634613932	
P(T<=t) one-tail	0.054802169	
t Critical one-tail	1.681951289	
P(T<=t) two-tail	0.109604338	
t Critical two-tail	2.018082341	

Because the test statistic (t Stat = 1.634613932) does not fall within the critical region (t Critical 2.018082341) we fail to reject the null hypothesis.

Note:

It is worth mentioning that this chart is not a "live" chart so that any changes made to the original data at this point would require using the Data Analysis Tool a second time to produce new test results.

DDXL – 2 VAR T TEST

Use the same data about **Discrimination Based on Age** with a column of data containing the ages of unsuccessful applicants and a column of data containing the ages of successful applicants.

1) Click on **DDXL**, select **Hypothesis Tests** and **2 Var t Test**.

2) In the dialog box click on the pencil icon for the **1ˢᵗ Quantitative Variable** and enter the range of data for the age of unsuccessful applicants as you did when using the t-Test: Two-Sample Assuming Unequal Variances in Excel. Then click on the pencil icon for the **2ⁿᵈ Quantitative Variable** and enter the range of data for the ages of successful applicants.

3) Click **OK**

4) Follow these steps in the **2 Sample t Test Setup** dialog box:

   a. **Step 1**: Select **2-sample.**

   b. **Step 2**: This step is optional so you can skip over it or set the difference at 0.

   c. **Step 3**: Set the significance level at 0.05.

   d. **Step 4**: Select the appropriate alternative hypothesis.

   e. **Step 5**: Click on **Compute**.

5) The results can be seen below. In addition to the Test Summary, DDXL also returns the mean and standard deviation of each variable and a conclusion to fail to reject the null hypothesis.

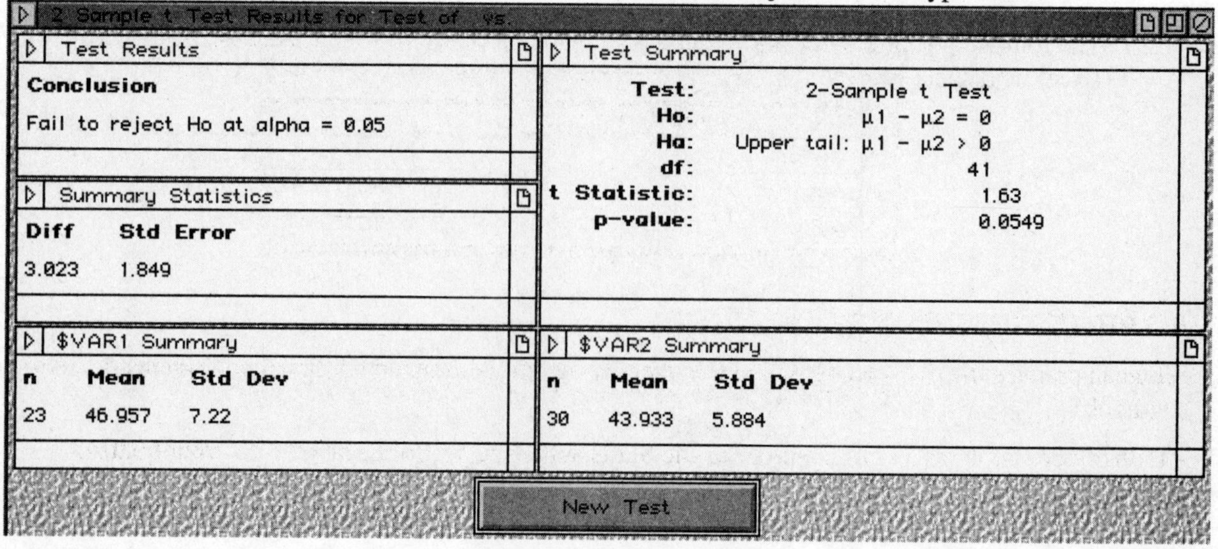

CONFIDENCE INTERVAL ESTIMATES

Using the same sample data concerning the age of unsuccessful and successful applicants we can construct a confidence interval estimate of the difference between the mean age of unsuccessful applicants and the mean age of successful applicants. This will be done using the DDXL add-in. The procedure for doing this is very similar to the one used to determine the 2 Var t Test.

1) Click on **DDXL**, select **Confidence Interval – 2 Var t Interval.**

2) In the dialog box click on the pencil icon for the **1ˢᵗ Quantitative Variable** and enter the range of data for the age of unsuccessful applicants. Then click on the pencil icon for the **2ⁿᵈ Quantitative Variable** and enter the range of data for the ages of successful applicants.

3) Click **OK**

4) In **2 Sample t Interval Setup**

     a. **Step 1:** Choose **2 sample**.

     b. **Step 2**: Select the appropriate confidence level (in this case 90%)

     c. **Step 3:** Click on **Compute Interval**.

5) The following results are displayed. In addition to the Confidence Interval results, DDXL also returns the mean and standard deviation for each variable.

```
┌ Confidence Interval ──────────────────────┐
│ ┌ Interval Results ──────────────────────┐ │
│ │ Confidence Interval                     │ │
│ │ With 90% Confidence, -0.0878 < μ1 - μ2 < 6.134 │ │
│ └────────────────────────────────────────┘ │
│ ┌ Interval Summary ──────────────────────┐ │
│ │ Diff    Std Err    df     t*            │ │
│ │ 3.023   1.849      41     1.682         │ │
│ └────────────────────────────────────────┘ │
│ ┌ $VAR1 Summary ─────────────────────────┐ │
│ │ n1    Mean                              │ │
│ │ 23    46.957                            │ │
│ └────────────────────────────────────────┘ │
│ ┌ $VAR2 Summary ─────────────────────────┐ │
│ │ n2    Mean                              │ │
│ │ 30    43.933                            │ │
│ └────────────────────────────────────────┘ │
└────────────────────────────────────────────┘
```

TO PRACTICE THESE SKILLS

You can practice the skills learned in this section by working through the following problems found in your textbook.

1) Work on problems 24 and 25 in the Basic Skills and Concepts for Section 9-3 of your textbook.

2) Use Data Set 14 found in Appendix B or in the data file COINS.xls to work through problem 27 in the Basic Skills and Concepts for Section 9-3 of your textbook.

SECTION 9-4: INFERENCES FROM MATCHED PAIRS

In the previous section we worked with independent populations. We now turn our attention to confidence interval estimates and hypothesis tests for dependent samples or matched pairs. The analysis can be done using either

- the Data Analysis **t-Test: Paired Two Samples for Means.**

- the DDXL add-in Hypothesis Tests: **Paired t Test**

Begin by entering the data found in Table 9–2 from the example **Hypothesis Test with Actual and Forecast Temperatures** into an Excel spreadsheet. This example is found in Section 9–4 of your textbook. Enter the data in columns rather than rows. The Data Analysis t Test requires that the data for each group be in a separate column. This is often referred to as **unstacked data.**

T-TEST: PAIRED TWO SAMPLES FOR MEANS

Using the data entered into Excel, we will test the claim that there is a difference between the actual low temperature and the low temperatures that were forecast five days earlier.

1) Click on **Tools,** highlight **Data Analysis** and click.

2) From the Analysis Tools list box in the Data Analysis dialog box select **t-Test: Paired Two Samples for Means.** Click **OK**.

3) In the t Test: Paired Two Samples for Means dialog box enter the following information

a. The cell range containing actual lows in the **Variable 1 Range** box.

b. The cell range for forecast lows from five days earlier in the **Variable 2 Range** box.

c. Enter 0.05 in the **Alpha** box. This value is supplied in the problem.

d. Determine where you wish to display the output.

e. Click **OK**.

4) The following summary of information containing calculations for the t-test is in added to your current Excel worksheet.

t-Test: Paired Two Sample for Means	Variable 1	Variable 2
Mean	57.4	58.4
Variance	19.8	11.3
Observations	5	5
Pearson Correlation	0.705311	
Hypothesized Mean Difference	0	
df	4	
t Stat	-0.70711	
P(T<=t) one-tail	0.259259	
t Critical one-tail	2.131846	
P(T<=t) two-tail	0.518519	
t Critical two-tail	2.776451	

In addition to the test statistic, Excel displays the P values for a one and two tailed test as well as the corresponding critical values. These values can be compared with those found in Step 6 of the solution for the example **Hypothesis Test with Actual and Forecast Temperatures.**

You can also use DDXL to perform the hypothesis test for this problem.

DDXL – PAIRED T TEST

1) Click on **DDXL** and choose **Hypotheses Tests** and **Pair t Test**.

2) In the Hypothesis test dialog box click on the pencil icon for the **1st Quantitative Variable** and enter the range of data for the actual lows. Click on the pencil icon for the **2nd Quantitative Variable** and enter the range of data for the lows from five days earlier. Click on the pencil icon for the **Pair Labels** and enter the range of data for the difference between the actual lows and forecast lows.

3) Click **OK**.

4) In a manner very similar to the previous hypotheses tests done with DDXL complete the fours steps in the **paired t test** dialog box.

5) Click on **Compute**.

6) The following results will be displayed.

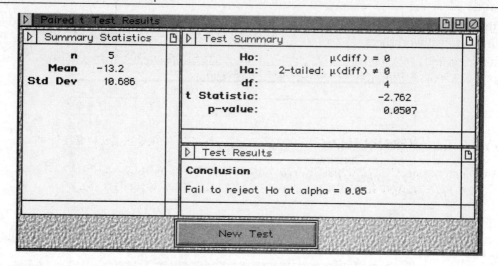

DDXL – CONFIDENCE INTERVALS FOR MATCHED PAIRS

Using the sample data from the preceding pages (Table 9–2) we can construct a confidence interval estimate of the mean of the differences between the actual low and the low temperatures from five days earlier. This will be done using the DDXL add-in. The procedure for doing this is very similar to the one used to determine the 2 Var t Test.

1) Click on **DDXL**, select **Confidence Interval – Paired t Interval**.

2) In the dialog box click on the pencil icon for the **1ˢᵗ Quantitative Variable** and enter the range of data for the actual lows. Click on the pencil icon for the **2ⁿᵈ Quantitative Variable** and enter the range of data for the lows five days earlier. Click on the pencil icon for the **Pair Labels** and enter the range of data for the difference between the actual lows and forecast lows.

3) Click **OK**.

4) Select the appropriate confidence level (in this case 95%) and click on **Compute Interval.**

5) The following results are displayed.

TO PRACTICE THESE SKILLS

Use the data presented in problems 13, 15 and 17 in the Basic Skills and Concepts for Section 9-4 of your textbook to practice the technology skills learned in this section.

SECTION 9-5: COMPARING VARIATION IN TWO SAMPLES

In this section we present a test of hypothesis for comparing two population variances. This will be done using the **F Test Two Sample for Variances** found in Excel's Data Analysis tools.

Open the file COLA.xls or enter the data found in Data Set 12 in Appendix B of your textbook. Determine the mean, variance and standard deviation using the weights of samples of regular Coke and regular Pepsi as seen below.

	regular coke	regular pepsi
mean	0.816822	0.824103
variance	0.000056	0.000033
standard deviation	0.0075074	0.0057011

Use this information and the **F Test Two Sample for Variance** found in Excel to work through the **Coke versus** Pepsi example found in Section 9-5 of your statistics book.

F-TEST: TWO SAMPLE FOR VARIANCES

Using the data entered into Excel, we will test the claim of equal population standard deviations.

1) Click on **Tools,** highlight **Data Analysis** and click.

2) From the Analysis Tools list box in the Data Analysis dialog box select **F-Test Two Sample for Variances.**

3) Click **OK**.

4) In the F- Test: Two Sample for Variances dialog box enter the following information:

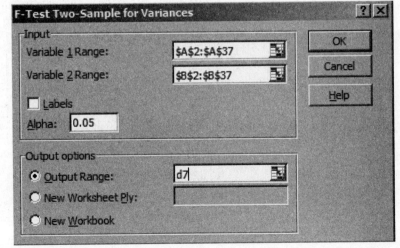

    a. The range of cells containing the weight of regular Coke samples in the **Variable 1 Range** box.

    b. The range of cells for the weight of regular Pepsi in the **Variable 2 Range** box.

    c. Enter 0.05 in the **Alpha** box. This value is supplied in the problem.

    d. Determine where you wish to display the output.

e. Click **OK**.

5) The following summary of information containing calculations for the F-test is in added to your current Excel worksheet.

F-Test Two-Sample for Variances		
	Variable 1	Variable 2
Mean	0.816822222	0.82410278
Variance	5.63606E-05	3.2502E-05
Observations	36	36
df	35	35
F	1.734067094	
P(F<=f) one-tail	0.054053516	
F Critical one-tail	1.757140211	

Excel will return the F test statistic, the P-value for the one-tailed case, and the critical F value for the one-tailed case. For a two-tailed test you can double the P-value returned by Excel.

TO PRACTICE THESE SKILLS

You can practice the technology skills learned in this section by working on the following problems found in your textbook.

1) Use Data Set 12 found in Appendix B or in the data file COLA.xls to work through problem 9 in the Basic Skills and Concepts for Section 9-5 of your textbook.

2) Use Data Set 14 found in Appendix B or in the data file COINS.xls to work through problem 15 in the Basic Skills and Concepts for Section 9-5 of your textbook.

3) Use Data Set 10 found in Appendix B or in the data file BOSTRAIN.xls to work through problem 19 in the Basic Skills and Concepts for Section 9-5 of your textbook.

CHAPTER 10: CORRELATION AND REGRESSION

SECTION 10-1: OVERVIEW 124

SECTION 10-2: CORRELATION 124

TO PRACTICE THESE SKILLS 125

SECTION 10-3 & 10-4: REGRESSION, VARIATION AND PREDICTION INTERVALS 126

TO PRACTICE THESE SKILLS 130

SECTION 10–5: MULTIPLE REGRESSION 130

TO PRACTICE THESE SKILLS 132

SECTION 10-6: MODELING 132

TO PRACTICE THESE SKILLS 136

SECTION 10-1: OVERVIEW

In this chapter we will be working with data that comes in pairs. We will be determining whether there is a relationship between the paired data, and will be trying to identify the relationship if it exists.

Excel provides an excellent tool to help us consider whether there is a statistically significant relationship between two variables. We can create a scatter plot, find a line of regression, and use our regression equation to predict values for one of the variables when we know values of the other variable.

The new functions introduced in this section are outlined below.

CORREL

This returns the correlation coefficient between two data sets. Your paired data must be entered in adjacent columns.

ADD TRENDLINE

This feature adds the linear regression graph to the scatter plot of a set of data values.

REGRESSION

This function returns information on Regression Statistics, as well as other information based on the linear regression equation. Your data values must be entered in adjacent columns.

SECTION 10-2: CORRELATION

In this section, we will take a look at a picture of a collection of paired sample data (duration in seconds of an Old Faithful eruption, Interval in minutes after the eruption) to help us determine if there appears to be a relationship between the variable x (duration of an eruption in seconds) and the variable y (interval in minutes after the eruption). We will work with a sample of the data from Data Set 11: Old Faithful Geyser in Appendix B in your book.

1) **Enter the data in Excel:** In a new worksheet, enter the following data. Make sure that you keep the pairs as they are listed.

2) **Compute the correlation coefficient:** In cell C1, type" r =", and move your cursor to cell D1. To find the linear correlation coefficient, click on the **Function** icon on the main menu bar, or click on **Insert, Function.** In the "Search for a function" box, type in correlation, and then click on **OK.** Click on **CORREL** in the Select a function box. Then click on **OK.**

	A	B
1	**Duration**	**Interval After**
2	240	92
3	120	65
4	178	72
5	234	94
6	235	83
7	269	94
8	255	101
9	220	87

a. In **Array1**, enter the range of cells where the data you want to use on your horizontal axis (x) is stored. In **Array2**, enter the range of cells where the data you want to use on your vertical axis (y) is

Function Arguments

CORREL

Array1 A2:A9   = {240;120;178;234;2

Array2 B2:B9   = {92;65;72;94;83;94

= 0.925591197

Returns the correlation coefficient between two data sets.

Array1 is a cell range of values. The values should be numbers, names, arrays, or references that contain numbers.

Formula result =   0.926

Help on this function   OK   Cancel

stored. Then click on **OK**. You will see the correlation coefficient in cell D1. This value should round to 0.926. This value indicates that there is a linear relationship between the height of the eruption and the time interval after an eruption.

3) **Create the scatter plot:** To create the scatter plot for this data, click on the **Chart Wizard** icon on the main menu, or click on **Insert, Chart.** Click on **XY(Scatter)**, and then click on **Next**.

    a. In the **Data Range** box, enter the range of cells where your paired data is stored. If you entered your data in columns A and B starting with cell A2, you could type in "A2:B9". Make sure that the bullet by **Columns** is marked. Then click on **Next**.

    b. Name your graph and your axes appropriately. You can also choose what other types of features you want to be included on your graph by accessing the tabs at the top of the **Chart Wizard** window. Click on **Next**.

    c. Decide whether to insert your chart as an object in your current sheet, or in another sheet. Then click on **Finish**.

    d. Make appropriate adjustments to your chart size, font size, etc. to create a reasonable picture. Remember you can right click while your cursor is in any part of your chart to access formatting options for that particular region. For example, notice that there are no values shown which have an x coordinate (duration) less than 120, and a y coordinate (Interval after) less than 65, so eliminating that part of the graph is recommended. You can do this by right clicking on each of the axes, and then adjusting the scale. We chose to use a minimum value of 100 for our horizontal axis, and a minimum value of 60 for our vertical axis. Your scatter plot should appear as the one shown below:

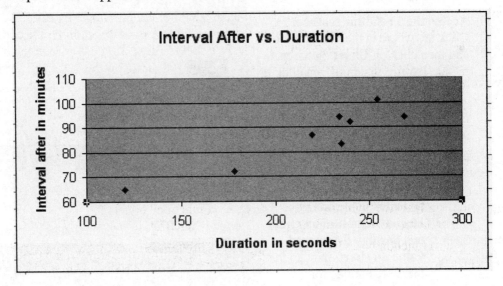

TO PRACTICE THESE SKILLS

You can apply the skills learned in this section by working on the following exercises.

1) Enter the data found in exercise 14 from Section 10-2 Basic Skills and Concepts in your textbook into Excel. Create the scatter plot, and find the linear correlation coefficient.

2) You can use Excel to work with exercises 15 through 28 from Section 10-2 Basic Skills and Concepts in your textbook. You need to enter the appropriate data given in the textbook.

3) You can use Excel to work with exercises 29 through 32 from section 10-2 Basic Skills and Concepts in your textbook. You can load the appropriate data from the CD that comes with your textbook, or find it at the website www.aw-bc.com/triola . After you have loaded the data, you should copy the appropriate columns to a new worksheet. Create the scatter plot and find the linear correlation coefficient.

SECTION 10-3 & 10-4: REGRESSION, VARIATION AND PREDICTION INTERVALS

In section 10-2 we concluded that there was evidence of linear correlation between the duration time of an eruption and the interval after the eruption. Now we want to determine this relationship in order to be able to calculate the interval time after an eruption once we know the duration of the eruption.

We have two options when working with linear regression, both of which are outlined below.

- The first option (**Add Trendline**) allows us to quickly generate the line of regression directly from our scatter plot.

- The second option uses the data analysis feature, and gives us a much more information, which will be useful in considering a more thorough analysis of the situation.

Option 1 – Add Trendline

1) **Access the Trendline feature:** Click anywhere in the **Chart** region, and then click on **Chart** on the main menu. (Notice that until you click in the Chart region, Chart is not an option in the main menu line.) Select **Add Trendline**. Make sure that **Linear** is selected from the possible types shown.

   a. Click on the **Options** tab, and click in the box by **Display equation on chart**, and **Display R-squared value on chart**. Notice that there are options for Forecasting forward and backwards. Initially, the Trendline will automatically only include the beginning and ending output values from your data set. To extend the line to fill up more of the graph you can use the "Forecast" feature. In our example, we can go forward by 20 units, and backward by 3 units. Click on **OK.** You will see your Trendline and the equation superimposed on your scatter plot.

2) **Modify your initial picture:** You will probably want to reformat your font size for the equation, and reposition where the equation appears on the screen so that it does not cover any of your data points.

    a. To modify your equation, you can select the region containing the equation, then right click and select the "Format Data Labels" option.

    b. To move the equations, select the region containing the equation and move your cursor in this region until an arrow appears. Hold down the left click button, and move the box where you want it to be within your plot area.

    c. You may find that when you used the forecast feature, your original scaling on your scatter plot changed. You can make additional changes if desired by double clicking on the axis that you want to adjust, selecting the tab labeled **Scale,** and making appropriate changes in the menu that appears.

3) **Final result:** Your picture should look similar to the one shown below. Notice that the equation shown does round off to the same equation that is shown in your book if you only wanted to use 3 significant digits for the slope and y intercept.

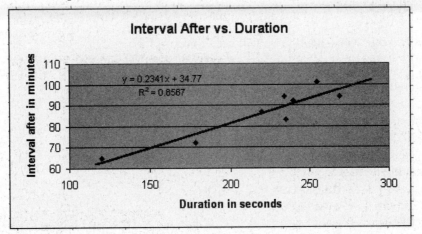

Option 2 – Data Analysis: Regression

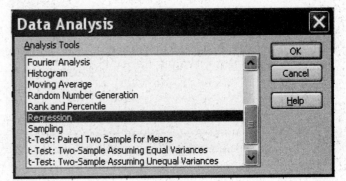

1) **Access the Regression Data Analysis Tool:** Click on **Tools** from the main menu, then click on **Data Analysis**, and click on **Regression.** Click on **OK.**

2) **Complete the Regression dialog box as shown.** The cell addresses shown assume that you entered your x value label in cell A1, and your x values (Duration times in seconds) in cells A2 through A9. Your y value label is in cell B1, and your y values (Interval After in Minutes) are in cells B2 through B9. If your data is entered in other cells, you should make appropriate adjustments.

a. **You should select both the column labels and the data in order to get the most useful, appropriately labeled output in terms of the contextual problem.** Notice that if you do include the column headings, you must check the **Labels** box. Otherwise Excel will give you an error message saying that you have selected non numeric data.

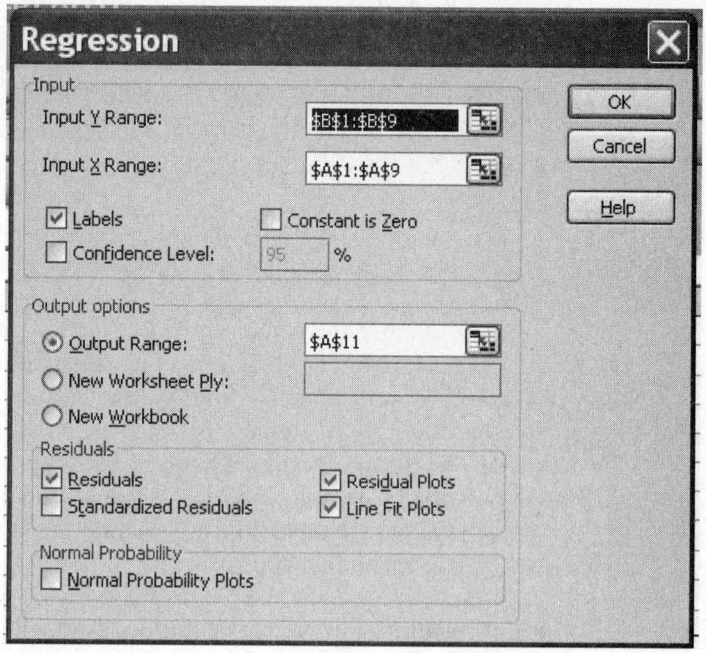

b. You can either type in the cell addresses with a colon in between the beginning and ending cell, or you can select the cells in your worksheet.

c. Notice, it is imperative that you are clear which data values represent your vertical (y) axis, and which data values represent your horizontal (x) axis.

3) **Click on OK**.

4) **Modify your worksheet:** You will need to resize the columns to see all the information provided clearly. Remember, to do this you can click on **Format,** click on **Columns,** and then click on **AutoFit Selection**. You will see the information shown below.

SUMMARY OUTPUT

Regression Statistics	
Multiple R	0.925591197
R Square	0.856719063
Adjusted R Square	0.832838907
Standard Error	4.973916136
Observations	8

ANOVA

	df	SS	MS	F	Significance F
Regression	1	887.5609496	887.560949	35.87577315	0.00097332
Residual	6	148.4390504	24.73984173		
Total	7	1036			

	Coefficients	Standard Error	t Stat	P-value	Lower 95%	Upper 95%	Lower 95.0%	Upper 95
Intercept	34.7698041	8.732045414	3.981862491	0.00726777	13.40325873	56.13634947	13.40325873	56.13634
Duration	0.234061432	0.039077721	5.989638817	0.00097332	0.138441695	0.329681169	0.138441695	0.329681

RESIDUAL OUTPUT

Observation	Predicted Interval After	Residuals
1	90.94454775	1.055452252
2	62.85717592	2.142824076
3	76.43273897	-4.432738972
4	89.54017916	4.459820843
5	89.77424059	-6.774240589
6	97.73232927	-3.732329272
7	94.45546923	6.544530774
8	86.26331911	0.736680889

5) **Modify your scatter plot:** You will also see a scatter plot showing both the actual data points as well as the points generated from the regression equation. You will need to resize and reformat this graph to make it look the way you want it to. You will also need to add the actual regression line. To add the line, double click on any one of the **Predicted** points. In the **Format Data Series** box that comes up, click in the bubble in front of **Automatic** under **Line**. (This can be found under the **Patterns** tab option.) You should create a picture similar to that shown below.

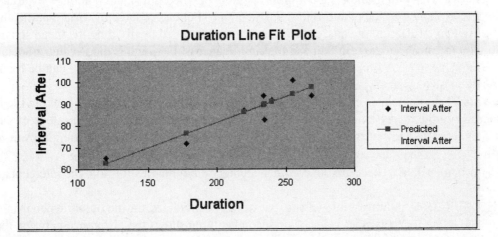

6) **Modify your residual plot:** Because we checked the box for Residual Plots, you will see a plot that shows your duration times along the horizontal axis, and which shows your residuals along the vertical axis. Your residual values are the differences between the actual data value and the predicted value created by using the regression equation. You should create a picture similar to that shown below.

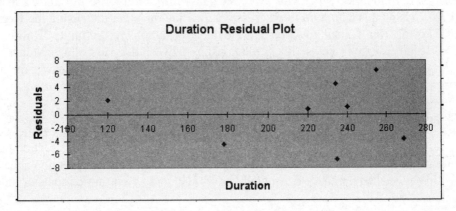

Interpreting This Output

Using the **Regression** option under **Data Analysis** provides you with more information than you need, but you can cut and paste the information that you need into another worksheet, or another document. To give you an idea of what the provided information represents, the major results are briefly described below:

- **Multiple R**: This is the correlation between the input variable (Duration time) and the output variable (Interval after). Since for this example there is only one input, the value given here is the correlation coefficient, r, expressing the linear relationship between the number of boats and the number of manatee deaths.
- **R Square**: This is also referred to as the coefficient of determination. It represents the proportion of variation in the output that can be explained by its linear relationship with the input.
- **Adjusted R Square**: The sample R Square tends to be an optimistic estimate of the fit between the model and the real population. The adjusted R Square gives a better estimate.
- **Standard Error**: This is the standard error of the estimate, and can be interpreted as the average error in predicting the output by using the regression equation.
- **Observations**: This is the number of paired data values included in the analysis.
- **ANOVA**: You do not need to understand most of the information provided in this section for this chapter, but essentially this part of the output gives more detailed information about the variation in the output that is explained by the relationship with the input. For each source of variation, the output gives degrees of freedom (df), sum of squares (SS), the F value obtained by dividing the mean square (MS) regression by the mean square residual, and the significance of F, which is the P-value associated with the obtained value of F. A fuller treatment of the Analysis of Variance (ANOVA) can be found in chapter 12.
- **Coefficients**: These are the coefficients for your regression equation. The value listed in the first row is the y intercept of the regression line, while the value listed in the second row is the slope of the line.
- **T Stat**: This refers to a test of the hypotheses that the intercept is significantly different from zero.
- **P – Value:** This is the probability associated with the obtained t statistic.
- **Lower and Upper 95%:** These are the confidence interval boundaries for both the intercept and the slope.
- **Residuals:** This table shows you the values that would be predicted for the output when using the regression equation for each input value. The second column shows the difference between the predicted value and the actual data value.

TO PRACTICE THESE SKILLS

You can practice the skills learned in this section by working on the following:

- Open one or more of the files where you saved the data from exercises 15 through 32 in Section 10-2 of your textbook. Using this data, find the regression lines and the other data using the **Regression** option for the paired data. Think about how you can use the information created under "Residual Output" to help you see whether there is a close linear relationship between the pairs of data values for each exercise. You can also use information generated in your tables to help you answer the questions asked for these exercises in Section 10-3 Basic Skills and Exercises in your textbook.

SECTION 10–5: MULTIPLE REGRESSION

The previous sections dealt with relationships between exactly two variables. This section presents a method for analyzing relationships that involve more than two variables. A multiple regression equation expresses a linear relationship between an output or dependent variable (y) and two or more inputs or independent variables (x values).

For this demonstration, we will use information found in table 10-1 of your text book on sample data from Old Faithful. We will consider our y variable as the time interval after the eruption, and our x variables as the duration time of the eruption and the height of the eruption.

	A	B	C
1	**Interval After**	**Duration**	**Height**
2	92	240	140
3	65	120	110
4	72	178	125
5	94	234	120
6	83	235	140
7	94	269	120
8	101	255	125
9	87	220	150
10			

1) **Enter the information**: Enter the given information into a new Excel worksheet. This data is taken directly from Table 10-1 in your text book. **The values for the independent x values must be in adjacent columns**.

2) **Fill out the Regression Dialog Box:** Click on **Tools, Data Analysis, Regression**. Assuming that you have entered your data into the same cells as shown for this example, you would fill in the **Regression** dialog box with Y range being A1 to A9, and X range being B1 to C9. Click on **OK**.

3) **You will see the information given below**. From this information, you can create the equation: Time Interval After = 45.1 + .245 (Duration) - 0.098(Height). For a discussion on other important elements, refer to the discussion in section 10-5 of your textbook.

Regression dialog box:

Input
- Input Y Range: $A$1:$A$9
- Input X Range: $B$1:$C$9
- ☑ Labels   ☐ Constant is Zero
- ☐ Confidence Level: 95 %

Output options
- ⦿ Output Range: $A$11
- ○ New Worksheet Ply:
- ○ New Workbook

Residuals
- ☐ Residuals   ☐ Residual Plots
- ☐ Standardized Residuals   ☑ Line Fit Plots

Normal Probability
- ☐ Normal Probability Plots

[OK] [Cancel] [Help]

SUMMARY OUTPUT

Regression Statistics	
Multiple R	0.93086
R Square	0.866501
Adjusted R Square	0.813101
Standard Error	5.259374
Observations	8

ANOVA

	df	SS	MS	F	Significance F
Regression	2	897.694939	448.847516	22672	0.006512
Residual	5	138.305060	327.66101		
Total	7	1036			

	Coefficients	Standard Error	t Stat	P-value	Lower 95%	Upper 95%	Lower 95.0%	Upper 95.0%
Intercept	45.10493	19.41148791	2.32362	0.067747	-4.79389	95.00375	-4.79389	95.00375
Duration	0.244636	0.044862	255.453057	0.002819	0.129314	0.359958	0.129314	0.359958
Height	-0.09825	0.162322111	-0.60528	0.571412	-0.51551	0.319012	-0.51551	0.319012

RESIDUAL OUTPUT

Observation	Predicted Interval After	Residuals
1	90.06263	1.937371636
2	63.65377	1.346232288
3	76.36893	-4.36892521
4	90.55982	3.440184113
5	88.83945	-5.839446295
6	99.12209	-5.122090372
7	95.20593	5.794070924
8	84.1874	2.812602916

TO PRACTICE THESE SKILLS

You can practice the skills learned in this section and in the previous section by working on exercises 13 through 16 from Section 10-5 Basic Skills and Concepts in your textbook.

SECTION 10-6: MODELING

We have used Excel to help us generate linear models for data sets. Although the **Regression** function does not give us the option to work with other types of models, we can generate scatter plots, and then add various types of Trend lines, including their equations. Your book shows various models that can be used with the TI-83 Plus calculator. In this section, we will use Excel to generate Quadratic and Exponential models for the data set given in Table 10-4 of your book.

1) **Enter the data:** Enter the data for the Coded Year and the Population into a new worksheet.

	A	B
1	Coded Year	Population
2	1	5
3	2	10
4	3	17
5	4	31
6	5	50
7	6	76
8	7	106
9	8	132
10	9	179
11	10	227
12	11	281
13		

2) **Create the scatter plot:** To create the scatter plot for this data, click on the **Chart Wizard** icon on the main menu, or click on **Insert, Chart.** Click on **XY (Scatter)**, and then click on **Next.**

   a. In the **Data Range** box, enter the range of cells where your paired data is stored. If you entered your data in columns A and B starting with cell A2, you could type in "A2:B12". Make sure that the bullet by **Columns** is marked. Then click on **Next.**

   b. Name your graph and your axes appropriately. You can also choose what other types of features you want to be included on your graph by accessing the tabs at the top of the **Chart Wizard** window. Click on **Next.**

   c. Decide where you want your chart to appear. Click on **Finish.**

   d. Make appropriate adjustments to your chart size, font size, etc. to create a reasonable picture. Remember you can right click while your cursor is in any part of your chart to access formatting options for that particular region. Your scatter plot should appear as the one shown below:

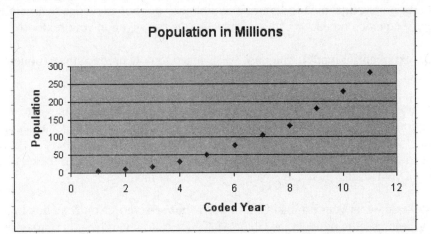

3) **Add the Trendline:** Click anywhere in the **Chart** region, and then click on **Chart** on the main menu. (Notice that until you click in the Chart region, Chart is not an option in the main menu line.) Select **Add Trendline**.

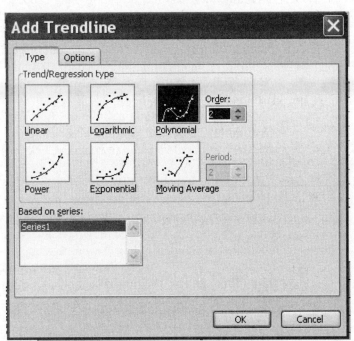

a. We will first find the Quadratic Model, so make sure that you have selected **Polynomial,** and that the **Order** is set at 2.

b. Click on the **Options** tab, and click in the box by **Display equation on chart**, and **Display R-squared value on chart**.

c. Click on **OK.** You will see your Trendline and the equation superimposed on your scatter plot.

4) **Make changes to your graph:** You will probably want to reformat your font size for the equation, and reposition where the equation appears on the screen so that it does not cover any of your data points. You can do this by selecting the region containing the equation, then right clicking and selecting the "Format Data Labels" option. To move

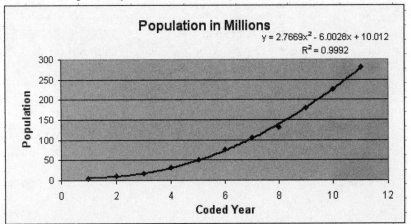

the equations, select the region containing the equation and move your cursor in this region until an arrow appears. Hold down the left click button, and move the box where you want it to be within your plot area. Your final picture should look like the one shown. Notice that with appropriate rounding, the equation produced by Excel matches the one given in your textbook.

5) **Create the exponential model:** Suppose you wanted to now produce the exponential model for this same data set.

   a. You should first select the entire Chart Area by first clicking somewhere within the white region around your actual graph. You should then see the entire Chart Area "selected", and can then copy and paste the graph to another area of your worksheet so that you preserve the quadratic model. Once you have done this, you are ready to clear the quadratic Trendline from your scatter plot, and create an exponential model.

   b. Make sure that your original Chart Area is **not selected**. You can click in any other cell in the worksheet to deselect the Chart Area. Then slowly move your cursor over your Trendline until you see the "yellow tag" appear which shows that you are pointing to the Trendline itself. Right click when you see this tag. You will see a box with 2 options: Format Trendline, or Clear. Select **Clear**. Your Trendline should then be removed from the picture, and you should be left with your original scatter plot.

   c. Again, make sure that your Chart Area is selected so that you see the **Chart** option in the main menu line at the top of the page. Select **Chart**, and then select **Add Trendline.** Choose the **Exponential** option in the Add Trendline Dialog box. Then click on the **Options** tab, and make sure that you have checked off the options to display the equation and the r- squared value on the chart. Click on **OK**.

   d. After making some modifications to the graph, you should have a picture similar to the one shown below. Notice that the equation is not exactly the same as the one produced by the TI-83 Plus calculator. The calculator uses the exponential form y = a * b^x, whereas Excel uses the form a * e^ x. You may have learned about the natural exponential function (e^x) in another math course. If you were to compute e raised to the numerical part of the power shown, you would find that the value produced is equivalent to the value for b shown in the TI-83 Plus screen. The two exponential models are equivalent ways to represent the equation that best fits the data.

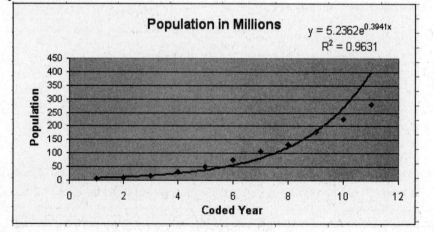

6) **Determine how close your model matches your data:** Let's suppose that you now wanted to find out how closely your model matched the data given. We will work with the Exponential model, but you can follow similar procedures for any of the models.

   a. First find the Chart Area which contains the Exponential model and equations in your Excel worksheet. We recommend writing down the equation on a piece of paper so that you don't have to keep referring to the graph.

   b. You need to input this expression in cell C2 in your worksheet, assuming that your first input value (coded year) is in cell A2 and your first output value (population in millions) is in cell B2. We need to modify this equation slightly so that Excel will understand that it is a formula. Make your modifications in the formula bar at the top of the screen. You need to insert the = sign before the expression so that Excel understands that it is a formula. You need to insert a * after the coefficient to indicate multiplication. To indicate that you want to use your natural exponential function, you would type EXP(, and then insert the exponent in parentheses. Your exponent needs to be adjusted by using a * after the numerical part, and then changing the x to a cell reference, in this case, cell A2. Your final formula bar should look as follows:

   f_x =5.2362*EXP(0.3941*A2)

   c. Once you have this formula entered correctly, press **Enter**. You should now see the value 7.76554 in cell C2.

   d. You can now use the fill command to copy this formula down the rest of the column. Your table should look similar to the one below.

Coded year	Population	Predicted Values from Exponential Model
1	5	7.765540368
2	10	11.51667568
3	17	17.07979258
4	31	25.33016668
5	50	37.56587447
6	76	55.71202679
7	106	82.62365704
8	132	122.5349192
9	179	181.7252706
10	227	269.5074529
11	281	399.692717

   e. You can follow similar procedures to create the predicted values for your quadratic model. The formula you would need to input for the quadratic model is shown below.

   Quadratic Model: f_x =2.7669*A2^2 - 6.0028*A2 + 10.012

f. You could now add other values into your Coded Year column, and continue to copy the formula for the predicted value down to include those input values.

TO PRACTICE THESE SKILLS

You can practice these skills by working on exercises 5 through 12 from Section 10-6 Basic Skills and Concepts in your textbook.

CHAPTER 11: MULTINOMIAL EXPERIMENTS AND CONTINGENCY TABLES

SECTION 11-1:   OVERVIEW **138**

SECTION 11 - 2:   MULTINOMIAL EXPERIMENTS:GOODNESS-OF-FIT **138**

SECTION 11-3:   CONTIGENCY TABLES: INDEPENDENCE AND HOMOGENEITY... 138

CREATING A TABLE OF EXPECTED FREQUENCIES **139**

PERFORMING THE CHI-SQUARE TEST **139**

TO PRACTICE THESE SKILLS **140**

SECTION 11-1: OVERVIEW

In earlier chapters you learned that the first step in organizing and summarizing data for a single variable was to create a frequency table. It is often advantageous to categorize data and create frequency counts for different variables. It is also desirable to label data according to two quantitative variables for the purpose of determining whether or not these variables are related. This data is organized by using a **contingency table** (or two-way frequency tables). You have already learned the basics for creating tables in Excel by using the **Pivot Table** wizard introduced in Chapter 4. We will look at the Chi Square test for Independence, used to determine whether a contingency table's row variable is independent of its column variable.

CHITEST: returns the test for independence: the value from the chi-squared distribution for the statistic and the appropriate degrees of freedom.

CHITEST (actual_range, expected_range) where the actual_range is the range of data that contains observations to test against expected values and the expected_range is the range of data that contains the ratio of the product of row totals and column totals to the grand total.

SECTION 11 - 2: MULTINOMIAL EXPERIMENTS: GOODNESS-OF-FIT

In previous chapters you looked at several different hypothesis tests that assumed that the data came from a normally distributed population. There are other less formal ways to check to see if a population is normally distributed. These might include creating a histogram and observing if the shape resembles a normal distribution. While this is not a bad approach, we will feel more secure about decisions we make if we can substantiate our findings with a formal statistical technique. A **goodness-of-fit test** is one such technique.

Excel does not contain a built in function that will perform a goodness-to-fit test. Much of the work presented in this section can be done using the traditional paper and pencil approach. However, the use of technology makes the computational aspect of these problems much easier. By entering a formula in a cell and then copying that formula throughout the appropriate cells, it is possible to save time and avoid arithmetic mistakes. Keep this in mind when working through problems for Section 11-2 in your statistics textbook.

SECTION 10-3: CONTIGENCY TABLES: INDEPENDENCE AND HOMOGENEITY

Excel provides the tools necessary to do a chi-square test for independence, although they are not found within a single analysis tool. The process involves (1) creating a contingency table using the **Pivot table** command in Excel, (2) determining the observed frequencies, (3) determining the expected frequencies and (4) using the **CHITEST** function.

The contingency table is the backbone of the chi-square test for independence in Excel.

Enter the information presented in Table 11-5 into Excel. This table represents the motorcycle helmets color and the number of riders wearing the helmet who were not injured (control) or injured or killed (cases).

	Black	White	Yellow/Orange	Total
1	Black	White	Yellow/Orange	**Total**
2 Controls	491	377	31	899
3 Cases	213	112	8	333
4 Total	704	489	39	1232

CREATING A TABLE OF EXPECTED FREQUENCIES

The contingency table provides the actual frequencies for each cell. To perform the chi-squared test for independence we also need the expected frequencies. Excel will expect to find these expected frequencies in a separate table and not within the table you just created.

1) Begin by copying the row headings (Controls, Cases) and column headings (Black, White, Yellow/Orange) to a different location within the same Excel worksheet. We will use this new table for the expected frequencies.

2) Find the **expected frequency** for each cell in the new table created in step (1), using the formula
$$\text{expected frequencies} = \frac{(row\ total)(column\ total)}{grand\ total}$$ and the appropriate cell addresses.

Note:

You must use relative and absolute references if you are going to copy your formula to all of your cells and not just retype them in. If you have forgotten how to use relative and absolute references see Chapter 1, Section 1-6, for help.

3) If all is done properly you should see

6	Black	White	Yellow/Orange
7 Controls	513.714	356.827	28.459
8 Cases	190.286	132.173	10.541

PERFORMING THE CHI-SQUARE TEST

To test the hypothesis presented in the **Injuries and Color of Motorcycle Helmet** example found in Section 11-3 of your textbook we will use the statistical function **CHITEST** and perform the chi-square test. This function asks for the observed and expected values and will return the p value of the test.

1) Click on **Insert**, highlight **Function** and click.

2) From the **Insert Function Dialog box** highlight **Statistical** in the category area and then highlight **CHITEST.**

3) Click **OK.**

4) The **CHITEST** dialog box opens.

    a. For the **Actual_range** highlight those cells that contain the observed frequencies (found in the first table we created). Be careful not to highlight the row and column totals.

    b. For the **Expected_range** highlight those cells that contain the expected frequencies (found in the second table we created).

5) Click **OK.**

This returns a very small P-value (P = 0.012434). Since the P-value is less than the significance level of 0.05 we reject the null hypothesis of independence between group and helmet color. It appears that helmet color and group (control or case) are dependent. It appears there is an association between helmet color and motorcycle safety.

TO PRACTICE THESE SKILLS

You can practice the skills presented in this section by working through problems 7, 9 and 13 found in the Basic Skills and Concepts for Section 11-3 of your textbook.

CHAPTER 12: ANALYSIS OF VARIANCE

SECTION 12-1: OVERVIEW 142

SECTION 12–2: ONE-WAY ANOVA 142

    TO PRACTICE THESE SKILLS 143

SECTION 12-3: TWO WAY ANOVA 143

    TO PRACTICE THESE SKILLS 145

SECTION 12-1: OVERVIEW

In this chapter, we consider a procedure for testing the hypothesis that three or more means are equal. We will use the Analysis of Variance (ANOVA) features of Excel.

Excel tools introduced in this section are outlined below.

ANOVA SINGLE FACTOR
This feature returns summary statistics on the data, as well as Analysis of Variance information.

ANOVA: TWO FACTORS WITH REPLICATION
This feature returns summary statistics for each group in your data set, as well as Analysis of Variance information.

SECTION 12–2: ONE-WAY ANOVA

1) **Enter the data in a worksheet:** We will use the data from table 12-1 in your text book. Type the data in as shown.

	A	B	C	D
1	Weights (kg) of Poplar Trees			
2		Treatment		
3	None	Fertilizer	Irrigation	Fertilizer and Irrigation
4	0.15	1.34	0.23	2.03
5	0.02	0.14	0.04	0.27
6	0.16	0.02	0.34	0.92
7	0.37	0.08	0.16	1.07
8	0.22	0.08	0.05	2.38
9				
10				

2) **Access ANOVA:** Click on **Tools** and click on **Data Analysis.** Click on **ANOVA: Single Factor**, and click on **OK.**

3) **Fill in the Dialog Box:** In the dialog box, type in "$A$3:$D$8" in the **Input Range,** or select these cells in your worksheet. Make sure that **Columns** and **Labels in First Row** are selected, that you have a check in the box by "Labels in First Row" and that **Alpha** is set at 0.05. You should have your results appear in a new worksheet. Then click on **OK.**

4) **Format the output:** You will see a table of values that is automatically selected.

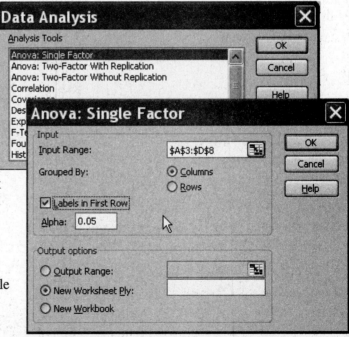

Click on **Format, Column**, and click on **AutoFit Selection**. You will see the table below. For a discussion of the key components of this information, read through the material presented in section 12-2 of your textbook.

Anova: Single Factor

SUMMARY

Groups	Count	Sum	Average	Variance
None	5	0.92	0.184	0.01613
Fertilizer	5	1.66	0.332	0.31932
Irrigation	5	0.82	0.164	0.01593
Fertilizer and Irrigation	5	6.67	1.334	0.73793

ANOVA

Source of Variation	SS	df	MS	F	P-value	F crit
Between Groups	4.682415	3	1.560805	5.731352875	0.007348294	3.238871522
Within Groups	4.35724	16	0.2723275			
Total	9.039655	19				

TO PRACTICE THESE SKILLS

You can practice the skills learned in this section by working on exercises 11 through 16 from Section 12-2 Basic Skills and Concepts in your textbook.

SECTION 12-3: TWO WAY ANOVA

For this demonstration, we will use the information presented in Table 12–4 of your textbook. You **MUST** enter the information for each type of treatment down a column, not across a row, and each row must begin with the category of Site 1 or Site 2.

1) **Enter the data in Excel**: Set up the table as shown.

2) **Access the ANOVA: Two Factor Replication**: Click on **Tools, Data Analysis,** and then click on **ANOVA: Two-Factor with Replication**. Click on **OK**.

3) **Fill out the dialog box**: Type in "$A$3:$E$13" in the box for **Input Range,** or select these cells in your worksheet. Type in 5 for **Rows per sample,** since you have 5 values for Site 1 and 5 values for Site 2 for each of the treatment groups. Type

	A	B	C	D	E
1			Weights (kg) of Poplar Trees		
2				Treatment	
3		None	Fertilizer	Irrigation	Fertilizer and Irrigation
4	Site 1	0.15	1.34	0.23	2.03
5	Site 1	0.02	0.14	0.04	0.27
6	Site 1	0.16	0.02	0.34	0.92
7	Site 1	0.37	0.08	0.16	1.07
8	Site 1	0.22	0.08	0.05	2.38
9	Site 2	0.6	1.16	0.65	0.22
10	Site 2	1.11	0.93	0.08	2.13
11	Site 2	0.07	0.3	0.62	2.33
12	Site 2	0.07	0.59	0.01	1.74
13	Site 2	0.44	0.17	0.03	0.12
14					

in 0.05 for **Alpha**. You should have your results appear in a new worksheet. Click on **OK**.

Anova: Two-Factor With Replication ☒

Input

Input Range: `$A$3:$E$13` ▦

Rows per sample: `5`

Alpha: `0.05`

Output options

◯ Output Range: ▦

◉ New Worksheet Ply:

◯ New Workbook

OK

Cancel

Help

4) **Format the results:** You will need to format the columns so that the titles all show up fully. You will see the data that is presented on the next page. For a detailed discussion of the values in this table, see section 12-3 in your textbook.

Anova: Two-Factor With Replication

SUMMARY	None	Fertilizer	Irrigation	Fertilizer and Irrigation	Total
Site 1					
Count	5	5	5	5	20
Sum	0.92	1.66	0.82	6.67	10.07
Average	0.184	0.332	0.164	1.334	0.5035
Variance	0.01613	0.31932	0.01593	0.73793	0.4757713
Site 2					
Count	5	5	5	5	20
Sum	2.29	3.15	1.39	6.54	13.37
Average	0.458	0.63	0.278	1.308	0.6685
Variance	0.18667	0.17325	0.10697	1.12547	0.4950345
Total					
Count	10	10	10	10	
Sum	3.21	4.81	2.21	13.21	
Average	0.321	0.481	0.221	1.321	
Variance	0.1109878	0.2435878	0.0582322	0.828365556	

ANOVA

Source of Variation	SS	df	MS	F	P-value	F crit
Sample	0.27225	1	0.27225	0.81218047	0.3742095	4.14909741
Columns	7.547	3	2.5156667	7.504776253	0.0006136	2.90111959
Interaction	0.17163	3	0.05721	0.170669769	0.915411	2.90111959
Within	10.72668	32	0.3352088			
Total	18.71756	39				

TO PRACTICE THESE SKILLS

You can practice the skills learned in this section by completing exercises 13 and 14 from Section 12-3 Basic Skills and Concepts in your textbook. Remember, when working with exercise 13, you **MUST** enter the data for each age group in columns, with male or female beginning each row.

CHAPTER 13: NONPARAMETRIC STATISTICS

SECTION 13-1: OVERVIEW 147

SECTION 13-2: SIGN TEST 147

USING EXCEL TO WORK WITH THE SIGN TEST 147

USING DDXL TO WORK WITH SIGN TEST 149

TO PRACTICE THESE SKILLS 150

SECTION 13-3 WILCOXON SIGNED-RANKS TEST FOR MATCHED PAIRS 151

USING EXCEL TO WORK WITH THE WILCOXON SIGNED – RANK TEST 151

USING DDXL TO WORK WITH THE WILCOXON SIGNED RANK TEST 153

TO PRACTICE THESE SKILLS 154

SECTION 13-4: WILCOXON RANK-SUM TEST FOR TWO INDEPENDENT SAMPLES

.... 154

TO PRACTICE THESE SKILLS 156

SECTION 13-5: KRUSKAL-WALLIS TEST 157

USING EXCEL FOR THE KRUSKAL WALLIS TEST 157

USING DDXL AND THE KRUSKAL- WALLIS TEST 157

TO PRACTICE THESE SKILLS 158

SECTION 13-6: RANK CORRELATION 158

TO PRACTICE THESE SKILLS 159

SECTION 13-1: OVERVIEW

In this chapter, many of the non-parametric methods used are not immediately supported by Excel. There are, however, some places where Excel can be used to help generate some of the intermediate steps in processes outlined in the sections.

The new functions we will use in this section are outlined below.

SIGN
This function determines the sign of a number. It returns 1 if the number is positive, zero (0) if the number is 0, and -1 if the number is negative.

COUNTA
This function counts the number of cells that are not empty and the values within the list of arguments. Use COUNTA to count the number of cells that contain data in a range or array.

COUNTIF
This function counts the number of cells within a range that meet the given criteria.

ABS
This function returns the absolute value of a number. The absolute value of a number is the number without its sign.

RANK
This function returns the rank of a number in a list of numbers. The rank of a number is its size relative to other values in a list. (If you were to sort the list, the rank of the number would be its position.)

SUMIF
This function adds the cells specified by a given criteria.

SECTION 13-2: SIGN TEST

USING EXCEL TO WORK WITH THE SIGN TEST

We will use Excel where possible to help us organize the information that we need to use in conducting the Sign Test. The steps outlined below would be particularly beneficial if you were working with a large data set. With a small data set, you can probably do all the work as quickly by hand, however continuing to use Excel when possible allows you to continue building proficiency with the program.

1) **Enter the data from Table 13-3:** The data shown is from your textbook. You should enter the data in columns as shown.

2) **Use the Sign Test:** To use the Sign Test, we want to convert the raw data to plus and minus signs. We can use Excel to help us with this process by first creating a formula which subtracts the pairs, then using the **SIGN** function to give us a column representing the sign of the difference, and finally using the **COUNTA** function to determine how many of the values are

	A	B	C	D
1	Yields of Corn from Different Seeds			
2	Regular	Kiln Dried		
3	1903	2009		
4	1935	1915		
5	1910	2011		
6	2498	2463		
7	2108	2180		
8	1961	1925		
9	2060	2122		
10	1444	1482		
11	1612	1542		
12	1316	1443		
13	1511	1535		
14				

positive or negative. To accomplish this, follow the steps below.

a. In cell C3, enter the formula: =A3-B3. This will take the first trial value and subtract the second trial value from it. Then use the fill command to copy this formula down the rest of the column.

b. In cell D3, click on **Insert, Function.** In the Search for a Function box, type Sign, and then press **Enter.** Make sure you have SIGN highlighted in the Select a function box, and then click on **OK.**

Insert Function ? X

Search for a function:

| sign | | Go |

Or select a category: Recommended ▾

Select a function:

SIGN
ABS
DEC2BIN
CEILING
HEX2DEC
BIN2HEX
EVEN

SIGN(number)
Returns the sign of a number: 1 if the number is positive, zero if the number is zero, or -1 if the number is negative.

Help on this function   OK   Cancel

c. In the dialog box that comes up, in the box after **Number,** type in C3 to indicate that the number you want to determine the sign for is in cell C3. Note that if the number is positive, the value returned will be 1. If the number is zero, the value returned will be 0. If the number is negative, the value returned will be -1. Click on **OK.**

Function Arguments X

SIGN

    Number c3   [📊] = -106

                  = -1

Returns the sign of a number: 1 if the number is positive, zero if the number is zero, or -1 if the number is negative.

        Number is any real number.

Formula result =   -1

Help on this function   OK   Cancel

d. Using the fill handle, copy this formula down the rest of the column. Your worksheet should now look similar to the one shown.

3) **Count the signs:** We now want to create a count of how many positive and how many negative values are contained in the sign column. Because this data set is small, it is easy to count the values, but we will still show how you can accomplish this with Excel so that you have tools which would make your work with a larger data set easier.

a. In cell E2, type Positive, and in cell E3 type Negative.

	A	B	C	D
1	Yields of Corn from Different Seeds			
2	Regular	Kiln Dried		
3	1903	2009	-106	-1
4	1935	1915	20	1
5	1910	2011	-101	-1
6	2498	2463	35	1
7	2108	2180	-72	-1
8	1961	1925	36	1
9	2060	2122	-62	-1
10	1444	1482	-38	-1
11	1612	1542	70	1
12	1316	1443	-127	-1
13	1511	1535	-24	-1
14				

b. Move to cell F2, and either click on the **Function** icon, or select **Insert, Function.** In the Search for a function box, type Count, and then press **Enter.** In the Select a function box, click on **Countif.** Then click on **OK.**

c. In the **Range** box, either enter $D$3:$D$13. These are the cells where our **SIGN** values are located in the worksheet. Since we will be copying this formula for the next count, it is essential that you use an absolute address.

d. Since we want to first find how many positive values there are, and since the positive values are indicated by 1, enter 1 in the **Criteria** box. Then click on **OK.** You should see the value 4 returned in cell F2.

e. Copy this formula to F3, and in the formula bar, insert a negative sign before the 1, so that your count will count the number of negative values in your range cells. Notice that the cell address for the Range stayed the same because you had used absolute cell addresses. Then press **Enter.** You should see the value 7 returned in cell F3.

4) **Use the BINOMDIST function to find the P-value:** Excel does not have a built- in function dedicated to the sign test, but after you have created the count from above, you can use the BINOMDIST function to find the P-value for the sign test. To do this, we will **use the number of times the LESS frequent sign occurs.** Access the BINOMDIST function, and fill in the dialog box as shown for this example. Since this is a two tailed test, we would double

Function Arguments		X
BINOMDIST		
Number_s	4	= 4
Trials	11	= 11
Probability_s	.5	= 0.5
Cumulative	True	= TRUE
		= 0.274414063

Returns the individual term binomial distribution probability.

Cumulative is a logical value: for the cumulative distribution function, use TRUE; for the probability mass function, use FALSE.

Formula result =   0.274414063

Help on this function   OK   Cancel

our result to get .548828. Since this is greater than the significance level of .05 for this example, we would fail to reject the null hypothesis.

USING DDXL TO WORK WITH SIGN TEST

We can use the Add-In DDXL to work with the Sign Test.
1) **Enter your data into Excel:** To work with the data from table 13-4 in your book, you need to enter it in Excel as shown.

2) **Access DDXL:** From the menu bar, click on **DDXL**, and then click on Nonparametric tests. Press the down arrow to the left of the Function type box, and click on **Paired sign test.**

	A	B
1	**Regular**	**Kiln Dried**
2	1935	1915
3	1511	1535
4	2496	2463
5	1961	1925
6	1444	1482
7	2060	2122
8	1612	1542
9	2108	2180
10	1910	2011
11	1903	2009
12	1316	1443
13		

a. Click on the Pencil Icon 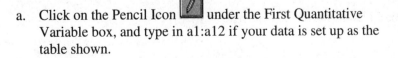 under the First Quantitative Variable box, and type in a1:a12 if your data is set up as the table shown.

b. Click on the Pencil Icon under the Second Quantitative Variable box, and type in b1:b12.

c. Click on **OK.** You will be taken to a screen like the one shown.

Data Desk® 6.1 Viewer - Untitled

File Edit Data Special Help

Paired Sign Test Setup

Step 1: Set the significance (alpha) level.

| 0.01 | 0.05 | 0.10 | Other... |

Step 2: Select an alternative hypothesis (Ha).

| Left Tailed | Two Tailed | Right Tailed |

Settings for Test of vs.

Alpha: 0.05
  Ho: Median (Var1 - Var2) = 0
  Ha: Select 1- or 2-tailed test.

Step 3: Compute.

d. Under **Step 1**, click on .05, since that is the significance level for the example worked out in your text book.

e. Under **Step 2,** click on Two Tailed.

f. Under **Step 3,** click on Compute. You should now see the results shown.

Test Results for Test of vs.

Test Summary

Ho:	Median (Var1 - Var2) = 0
Ha:	2-tailed: Median (Var1 - Var2) ≠ 0
Count	11
Count (Ties Adjusted)	11
Positive Diffs	4
Negative Diffs	7
p-value:	0.548

Test Results

Conclusion

Fail to reject Ho at alpha = 0.05

New Test

TO PRACTICE THESE SKILLS

You can practice these skills by working on exercises 10, 11 and 12 from Section 13-2 Basic Skills and Concepts in your textbook. You would enter the data for the data in adjacent columns. After creating your number of positive and negative values, follow the procedures outlined in your textbook to complete the hypothesis test.

SECTION 13-3 WILCOXON SIGNED-RANKS TEST FOR MATCHED PAIRS

Again, although Excel is not programmed for the Wilcoxon signed-ranks test, we can use Excel to help us complete some of the intermediate steps, particularly if the data set is large. We can also use the DDXL Add-In to work with this test.

USING EXCEL TO WORK WITH THE WILCOXON SIGNED – RANK TEST

We will use the data from Table 13-4 in your textbook to show you how and where Excel could be used to help produce information needed in order to perform the Wilcoxon signed-ranks test.

1) **Compute the differences:** Follow steps 1 – 2a from section 13-2 of this manual to produce the worksheet shown, or copy the appropriate columns if you saved your work from the previous section to a new worksheet.

2) **Delete any results of 0:** Step 2 in your textbook for the Wilcoxon Signed-Ranks Procedure tells us to discard any pairs for which the difference is = 0. The easiest way to proceed is to select any row where the difference is zero, select **Edit** from the main menu bar, and then select **Delete.** For the data shown, we do not have to delete any information.

	A	B	C	D
1	Yields of Corn from Different Seeds			
2	Regular	Kiln Dried		
3	1903	2009	-106	
4	1935	1915	20	
5	1910	2011	-101	
6	2498	2463	35	
7	2108	2180	-72	
8	1961	1925	36	
9	2060	2122	-62	
10	1444	1482	-38	
11	1612	1542	70	
12	1316	1443	-127	
13	1511	1535	-24	
14				

3) **Rank the absolute differences:** Step 2 of the procedure tells us to ignore the signs of the differences and sort the differences from lowest to highest.

   a. **Find the absolute values:** To quickly create a column where we only show positive values, we can access the **ABS** function in Excel. In the column next to the difference column, position your cursor next to the cell which shows the first difference. Select **Insert Function**, and in the Search for a function box, type in absolute value. In the Select a function box, click on **ABS**. Press **OK**. In the **Number** box, type in the cell where your first difference appears. Then click on **OK**. Use your fill command to copy this formula down the rest of the column.

   b. **Rank the absolute differences:** We now need to know the ranks of these absolute differences. In the adjacent column, position your cursor in the cell next to the first absolute value. Select **Insert, Function**, and in the Search for a function box, type Rank and press **Enter.** In the Select a function box, make sure **Rank** is highlighted and then

Function Arguments

RANK
Number D3 = 20
Ref $D$3:$D$13 = {20;24;33;36;38;62
Order 1 = TRUE

= 1

Returns the rank of a number in a list of numbers: its size relative to other values in the list.

Number is the number for which you want to find the rank.

Formula result = 1

Help on this function   OK   Cancel

click on **OK. Fill in the dialog box** as shown (or make adjustments where appropriate depending on where your data is located.) Make sure that the array shown in the **Ref** box is typed in using absolute addresses. In the **Order** box, you can type in any non zero number since the list of data is in ascending order. Then click on **OK**.

c. **Use the fill command** to copy this formula down the rest of the column. Your worksheet should now appear similar to the one shown. Notice that we didn't have any ranks which were the same, and therefore don't have to make any adjustments to the rank column. If there had been repeated ranks, we would have had to manually type in the mean of the ranks involved in the tie for each of those values.

	A	B	C	D	E
1	Yields of Corn from Different Seeds				
2	Regular	Kiln Dried	Differences	Abs Value	Rank
3	1903	2009	-106	106	10
4	1935	1915	20	20	1
5	1910	2011	-101	101	9
6	2496	2463	33	33	3
7	2108	2180	-72	72	8
8	1961	1925	36	36	4
9	2060	2122	-62	62	6
10	1444	1482	-38	38	5
11	1612	1542	70	70	7
12	1316	1443	-127	127	11
13	1511	1535	-24	24	2
14					

d. **Sort the data according to the signed differences:** We will need to associate each ranking with either a positive or negative value according to Step 3 of The easiest way to do this is to sort the Differences column in ascending order so that we can clearly see which absolute ranks are associated with negative values and which are associated with positive values. Select the cells containing the column headings and the values in the table (Cells a2:e13 in the worksheet shown), and then click on **Data** in the menu bar. Click on **Sort**. Fill out the Sort Dialog box as shown. Notice that when you click in the Header row circle, only the numerical data is highlighted. Click on **OK.** Your final table should look like the one shown underneath the Sort screen.

	A	B	C	D	E
1	Yields of Corn from Different Seeds				
2	Regular	Kiln Dried	Differences	Abs Value	Rank
3	1316	1443	-127	127	11
4	1903	2009	-106	106	10
5	1910	2011	-101	101	9
6	2108	2180	-72	72	8
7	2060	2122	-62	62	6
8	1444	1482	-38	38	5
9	1511	1535	-24	24	2
10	1935	1915	20	20	1
11	2496	2463	33	33	3
12	1961	1925	36	36	4
13	1612	1542	70	70	7
14					
15					
16					
17					

Sort

Sort by

Differences ⊙ Ascending ○ Descending

Then by

⊙ Ascending ○ Descending

Then by

⊙ Ascending ○ Descending

My data range has

⊙ Header row ○ No header row

Cancel

	A	B	C	D	E
1	Yields of Corn from Different Seeds				
2	Regular	Kiln Dried	Differences	Abs Value	Rank
3	1316	1443	-127	127	11
4	1903	2009	-106	106	10
5	1910	2011	-101	101	9
6	2108	2180	-72	72	8
7	2060	2122	-62	62	6
8	1444	1482	-38	38	5
9	1511	1535	-24	24	2
10	1935	1915	20	20	1
11	2496	2463	33	33	3
12	1961	1925	36	36	4
13	1612	1542	70	70	7
14					

4) **Find the sums:** In Step 4, you are told to find the sum of the absolute values of the negative ranks and the sum of the positive ranks. With the sorted table, it is easy to see that you want to add the ranks in cells E3 through E9, since they are associated with the negative differences in column C. Use your **SUM** function to find the appropriate sums. You should end up with the following results:

Sum of Abs. Value of Negative Ranks:	51
Sum of Positive Ranks:	15

5) **Follow the remaining steps in your text book:** From this point on, continue to follow the steps outlined in your textbook to complete the hypothesis test.

USING DDXL TO WORK WITH THE WILCOXON SIGNED RANK TEST

1) **Enter the data into Excel:** Make sure you have the paired data entered into Excel.

2) **Access DDXL:** In the menu bar of Excel, click on DDXL, and then click on Nonparametric tests. Click on the down arrow on the left hand side of the Function type box, and click on **Paired Wilcoxon.**

	A	B
1	Regular	Kiln Dried
2	1935	1915
3	1511	1535
4	2496	2463
5	1961	1925
6	1444	1482
7	2060	2122
8	1612	1542
9	2108	2180
10	1910	2011
11	1903	2009
12	1316	1443

a. **Fill in the Dialog Box:** In the dialog box, click on the Pencil Icon ![pencil] under the **1st Quantitative Variable** box, and enter A1:A12 into the box. Click on the Pencil Icon under the **2nd Quantitative Variable** box, and enter B1:B12 into the box. Click on **OK.** You will be taken to a screen like the one shown.

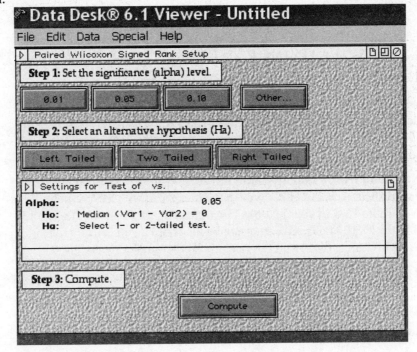

b. **Under Step 1:** Click on 0.05 for the significance level.

c. **Under Step 2:** Click on Two Tailed.

d. **Under Step 3:** Click on Compute. You will then see the screen below.

```
Data Desk® 6.1 Viewer - Untitled
File  Edit  Data  Special  Help
Test Results for Test of   vs.
  Test Summary

              Ho:            Median (Var1 - Var2) = 0
              Ha:    2-tailed: Median (Var1 - Var2) ≠ 0
           Count                              11
   Count Adjusted                             11
  Positive Ranks                              15
  Negative Ranks                              51
    Z Statistic:                            -1.6
        p-value:                           0.1095
      Conclusion       Fail to reject Ho at alpha = 0.05


                        New Test
```

TO PRACTICE THESE SKILLS

You can practice these skills by working on exercises 5 through 12 from Section 13-3 Basic Skills and Concepts in your textbook.

SECTION 13-4: WILCOXON RANK-SUM TEST FOR TWO INDEPENDENT SAMPLES

Although Excel is not programmed to compute the Wilcoxon Rank-Sum Test for Two Independent Samples directly, we can again make use of its features to help us obtain some of the information in the steps outlined in the procedure in section 13-4 of your textbook.

1) **Enter the data from Table 13-5 into Excel, or copy it from Data Set 1:** In a new worksheet, you can either type in the data for BMI Measurements shown in Table 13-5 in your text book or, since this data is from the first 13 values in the MHEALTH.XLS data set and the first 12 values from the FHEALTH.XLS data sets, you can open those files and copy and paste appropriate data into one worksheet.

2) **Combine the samples:** We need to create one large sample containing all of these values. In a new column, copy and paste the men's data, and then immediately after that, copy and paste the women's data.

Men	Women
23.8	19.6
23.2	23.8
24.6	19.6
26.2	29.1
23.5	25.2
24.5	21.4
21.5	22.0
31.4	27.5
26.4	33.5
22.7	20.6
27.8	29.9
28.1	17.7
25.2	

3) **Find the rank of each sample value in your combined list:** In the cell next to your first value of your combined data, insert the **Rank** function. **Make sure you use absolute cell references in your Ref box.** Your dialog box would look like the following if your combined sample was in column C of your worksheet, with the first value in cell C2. Click on **OK.** Then use the fill handle to copy the formula down the rest of the column. Your worksheet should now look like the one shown.

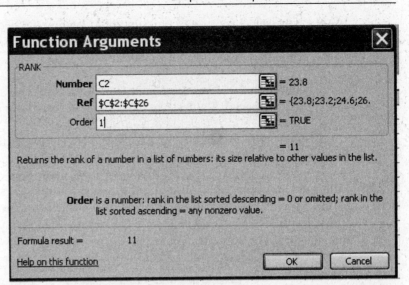

4) **Check for repeating ranks:** In order to see if there are any "repeating ranks", you should copy the rank column to another non-adjacent column in your worksheet, and then sort only that copied column. When you copy the column, you will need to use the **Paste Special** command, and paste the **Values** only. Since you want to be able to fill in the ranks for the values in each original column of data, you DO NOT want to rearrange the entire table of data! Notice that you see the rank 2, 11 and 15 twice. **In your original ranked column,** replace the 2's with 2.5 (the average of the 2nd and 3rd rankings), the 11's with 11.5 (the average of the 11th and 12th rankings) and the 15's with 15.5 (the average of the 15th and 16th rankings).

	A	B	C	D	E
1	Men	Women	Combined Data	Combined Rank	
2	23.8	19.6	23.8	11	
3	23.2	23.8	23.2	9	
4	24.6	19.6	24.6	14	
5	26.2	29.1	26.2	17	
6	23.5	25.2	23.5	10	
7	24.5	21.4	24.5	13	
8	21.5	22.0	21.5	6	
9	31.4	27.5	31.4	24	
10	26.4	33.5	26.4	18	
11	22.7	20.6	22.7	8	
12	27.8	29.9	27.8	20	
13	28.1	17.7	28.1	21	
14	25.2		25.2	15	
15			19.6	2	
16			23.8	11	
17			19.6	2	
18			29.1	22	
19			25.2	15	
20			21.4	5	
21			22.0	7	
22			27.5	19	
23			33.5	25	
24			20.6	4	
25			29.9	23	
26			17.7	1	

5) **Associate the ranks back with the original lists:** We want to associate the ranks with the original data values in our uncombined lists. You will need to insert 2 columns in your worksheet if it is set up like the one above.

a. Click on the B in the second column to select this column, and then click on **Insert** in the menu bar, followed by **Columns.** You should see a blank column in column B. Click on the column next to the Women's data (this should now be column D – Combined Data), and click on **Insert** followed by **Columns** again.

b. Now select the first 13 ranks in your Combined Ranks column. These are associated with the values in the Men's column. Copy and paste (**Use Paste Special, Values!**) them in the column next to the Men's column.

c. Select the last 12 ranks in your Combined Ranks column, and copy and paste them (**Use Paste Special, Values!**) in the column next to the Women's column.

d. Your table should now match Table 13-5 in your textbook, and should appear as the one shown below.

6) **Find the Count for each column and the sum of the Ranks:** You now want to compute the **Count** for each of your rank columns, as well as the **Sum**. In column A, type in Count and Sum below your data values. Then under the columns containing your ranked values, use your **Count** function and your **Sum** function to find the number of values in each column, as well as the sum of the rank values in each column. Your worksheet should look similar to the one shown.

	A	B	C	D
1	**Men**		**Women**	
2	23.8	11.5	19.6	2.5
3	23.2	9.0	23.8	11.5
4	24.6	14.0	19.6	2.5
5	26.2	17.0	29.1	22.0
6	23.5	10.0	25.2	15.5
7	24.5	13.0	21.4	5.0
8	21.5	6.0	22.0	7.0
9	31.4	24.0	27.5	19.0
10	26.4	18.0	33.5	25.0
11	22.7	8.0	20.6	4.0
12	27.8	20.0	29.9	23.0
13	28.1	21.0	17.7	1.0
14	25.2	15.5		
15				
16	**Count**	13		12
17	**Sum of Ranks**	187.0		138.0
18				

Although you can compute the mean and standard deviation of the sample *R* values using Excel, it is probably easier to do these computations using a calculator than to take the time entering the appropriate cell based formulas into Excel. Likewise, computing your test statistic is more simply done on a calculator.

Read the solution in your textbook to see how you use the values generated to make a decision in your hypothesis test.

TO PRACTICE THESE SKILLS

You can practice the skills learned in this section by working with exercises 7 through 12 from section 13-4 Basic Skills and Concepts in your textbook.

SECTION 13-5: KRUSKAL-WALLIS TEST

Again, Excel is not programmed to directly perform the Kruskal-Wallis Test. As in the previous section, you can use Excel to help you work through the preliminary work outlined in the procedures for performing the test in your textbook, or you can use the DDXL Add-In.

USING EXCEL FOR THE KRUSKAL WALLIS TEST

Since the only difference between this section and the previous section is that you are working with an additional column of data, we have chosen to just refer you back to section 13-4. You would be setting your initial data up in 3 columns. In a separate column, you would combine all of the data values, and then use the **Rank** function to find the Ranks for these values. Again, you would want to copy the ranked column, and order it so that it is easy to see if there are any repeated ranks. As before, you would replace the repeated ranks with the average value. Once you have changed the appropriate values in the original column of ranks, you would insert columns next to each of your original columns and then copy and paste (using Paste Special Values) the associated ranks in the columns next to the original data. You can then use your **Count** and **Sum** functions to find the values for n and R. At that point, you would use these values in the formula for H shown in your textbook.

USING DDXL AND THE KRUSKAL- WALLIS TEST

The Add-In DDXL can be used to find results from the Kruskal – Wallis Test.

1) **Enter your data into Excel:** For the example in the text book, you need to set up one column that contains the treatment types for the trees, and a second column that contains the weights. Your Excel worksheet should look like the one shown if you are using the data in Table 13-6 of your text book.

2) **Access DDXL:** Click on DDXL in the menu bar, and choose Nonparametric tests.
   a. Click on the down arrow to the left of the Function Type box, and click on Kruskal Wallis.

   b. Click on the Pencil Icon under the Response Variable box, and then type in B1:B20 if your spreadsheet is set up as the one shown.

   c. Click on the Pencil Icon under the Factor Variable box, and then type in A1:A20 if your spreadsheet is set up as the one shown.

	A	B
1	None	0.15
2	None	0.02
3	None	0.16
4	None	0.37
5	None	0.22
6	Fertilizer	1.34
7	Fertilizer	0.14
8	Fertilizer	0.02
9	Fertilizer	0.08
10	Fertilizer	0.08
11	Irrigation	0.23
12	Irrigation	0.04
13	Irrigation	0.34
14	Irrigation	0.16
15	Irrigation	0.05
16	Fert & Irr	2.03
17	Fert & Irr	0.27
18	Fert & Irr	0.92
19	Fert & Irr	1.07
20	Fert & Irr	2.38

   d. Click on **OK.** You should see the screen shown on the next page.

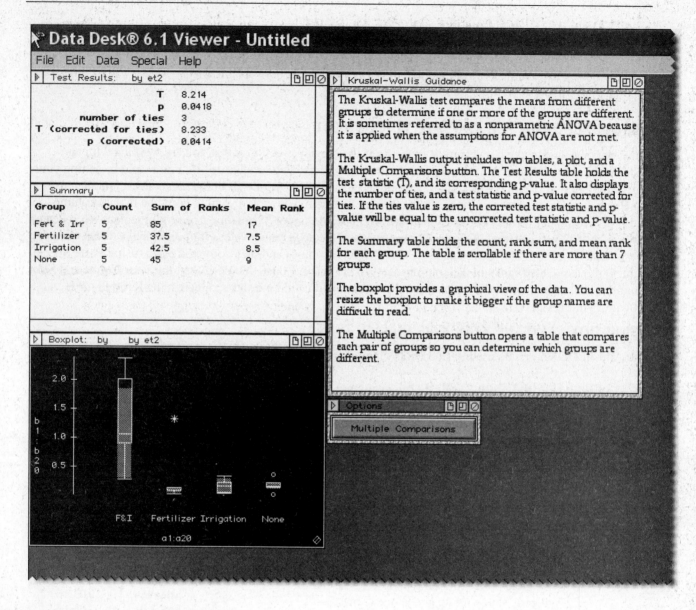

TO PRACTICE THESE SKILLS

To practice the skills outlined in this section, you can work on exercises 5 through 12 from Section 13-5 Basic Skills and Concepts in your textbook.

SECTION 13-6: RANK CORRELATION

We will use Excel to help us create the test statistic for the data shown in Table 13-7 of your textbook.

1) **Enter your Data in Excel:** In columns A and C of a new worksheet, type in the data for the College, Student Ranks and U.S. News and World Report Ranks.

	A	B	C
1	College	Student Ranks	U.S. N & WR Ranks
2	Harvard	1	1
3	Yale	2	2
4	Cal. Inst. Of Tech.	3	5
5	M.I.T.	4	4
6	Brown	5	7
7	Columbia	6	6
8	U. of Penn.	7	3
9	Notre Dame	8	8
10			

2) **Use the CORREL function:** Since this data is already given in terms of ranks, we just need to compute the rank correlation coefficient for the paired ranked data values.

   a. **In cell B 12:** Type in CORREL to indicate that the value in the next cell is the correlation coefficient.

   b. **In cell C12:** Insert the **CORREL** function, and fill in the dialog box so that **Array 1** refers to cells B1:B9 and **Array 2** refers to cells C1:C9. Press **Enter.** You should see a rank correlation coefficient of .71428. This matches the computation shown in your text book.

3) **If your data wasn't already ranked:** If you were given data that wasn't ranked already:

   a. You should set up your original data with a column in between, and then include the ranks for each original type of data in the column next to it.

   b. Use your **Rank** function to rank each column of data, making sure that your **Ref** box uses the absolute cell addresses for the range of cells where your data is located. Since you want your ranks to be computed for the data in ascending order, you should type in a non-zero value in the **Order** box.

   c. When you use your **CORREL** function, you would make reference to the cells containing the ranking values.

TO PRACTICE THESE SKILLS

You can practice the skills from this section by working on exercises 9 through 18 from Section 13-6 Basic Skills and Concepts in your textbook.

CHAPTER 14: STATISTICAL PROCESS CONTROL

SECTION 14–1: OVERVIEW **161**

SECTION 14–2: CONTROL CHARTS FOR VARIATION AND MEAN **161**

    TO PRACTICE THESE SKILLS **164**

SECTION 14-3: CONTROL CHARTS FOR ATTRIBUTES **164**

    TO PRACTICE THESE SKILLS **165**

SECTION 14–1: OVERVIEW

In this chapter, we address changing characteristics of data over time. In monitoring this characteristic, we are able to control the production of goods and services.

The major features used in this section are ones that have already been introduced in earlier sections, and include:

Chart Wizard to create a line graph from a set of data points.

Function used to access **Average, Median,** and **STDEV.**

The new feature introduced in this section is how to use the **Callouts** under the Draw menu.

SECTION 14–2: CONTROL CHARTS FOR VARIATION AND MEAN

In this section, we will consider data arranged according to some time sequence. We will consider the information on Aircraft Altimeter Errors (in feet) found in Table 14-1 in your textbook.

1) **Enter the data for the day and the Aircraft Altimeter Errors** in a new worksheet. It is not necessary to enter the data found in the last four columns of the table since this information can be determined using Excel.

2) **Determine** the **mean, median,** and sample **standard deviation** for each row using the function icon or **Insert, Function**.

Recall that the range is the difference between the highest and lowest value in the sample. This can be done in Excel using the MAX and MIN functions. For example, in the problem you are trying to work through – the range can be found by =MAX(B2:H2)-MIN(B2:H2).

3) To determine the **Range:**

   a. **Position** your cursor in cell H2

   b. **Enter the formula** = Max(B2:H2)-Min(B2:H2).

   c. Press **Enter.**

   d. **Copy the formula** down the rest of the column.

4) You should now have the table shown on the next page.

Day					Mean	Median	Range	St.Dev
1	2	-8	5	11	2.5	3.5	19	7.94
2	-5	2	6	8	2.8	4.0	13	5.74
3	6	7	-1	-8	1.0	2.5	15	6.98
4	-5	5	-5	6	0.3	0.0	11	6.08
5	9	3	-2	-2	2.0	0.5	11	5.23
6	16	-10	-1	-8	-0.8	-4.5	26	11.81
7	13	-8	-7	2	0.0	-2.5	21	9.76
8	-5	-4	2	8	0.3	-1.0	13	6.02
9	7	13	-2	-13	1.3	2.5	26	11.32
10	15	7	19	1	10.5	11.0	18	8.06
11	12	12	10	9	10.8	11.0	3	1.50
12	11	9	11	20	12.8	11.0	11	4.92
13	18	15	23	28	21.0	20.5	13	5.72
14	6	32	4	10	13.0	8.0	28	12.91
15	16	-13	-9	19	3.3	3.5	32	16.58
16	8	17	0	13	9.5	10.5	17	7.33
17	13	3	6	13	8.8	9.5	10	5.06
18	38	-5	-5	5	8.3	0.0	43	20.39
19	18	12	25	-6	12.3	15.0	31	13.28
20	-27	23	7	36	9.8	15.0	63	27.22

Control Chart for Monitoring Variation: The R Chart

Before we create the actual chart for the ranges, we need to compute the value for the Centerline, and the values for the Upper Control Limit (UCL) and the Lower Control Limit (LCL).

1) **Determine the mean for the sample ranges.** To do this, find the mean of the column containing the range values.

2) To find the upper and lower control limits, we must use Table 14–2 in your textbook. **Locate the value** for D_3 and D_4. In this case since $n = 4$ we find $D_3 = 0.000$ and $D_4 = 2.282$.

3) **Determine the upper control limit** using the formula $D_3 \overline{R}$. Your upper control limit will be 48.4

4) **Determine the lower control limit** using the formula $D_4 \overline{R}$. Your lower control limit will be 0.00.

5) **Create a scatterplot**, graphing $\overline{R}$ on the vertical and time (day) along the horizontal.

6) At this point you will need to determine the difference between the UCL and the mean of the range values, the LCL and the mean of the range values.

7) Work to set your horizontal gird lines so that they represent these three lines (UCL, LCL and $\overline{R}$)

8) Once you are satisfied that you accurately display the information required for the control chart, label the horizontal lines on your graph. This can be done using a textbox.

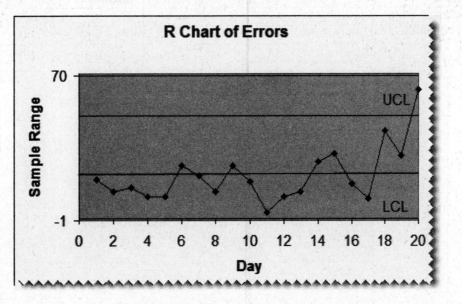

9) To add a textbox click on **Textbox** on the **Drawing Toolbar** at the bottom of your screen. (If the Drawing Toolbar is not available, click on **View,** select **Toolbars**, and from this menu, click in front of **Drawing.**)

Control Chart for Monitoring Means

You can follow basically the same procedures as those above to create a chart for the sample means. The major differences are outlined below.

1) Your **Chart Title** will need to reflect the fact that you are using the sample means, not the sample ranges.

2) To **find the upper and lower control limits**, we use Table 14–2 in your textbook and locate the value for A_2. In this case since $n = 4$ we find $A_2 = 0.729$.

3) **Find the mean of the sample means**. To do this, find the mean of the column containing the mean values. You should find that the mean of the sample means is 6.45.

4) **Determine the upper control limit** using the formula $\overline{\overline{X}} + A_2\overline{R}$. Your upper control limit will be 21.9.

5) **Determine the lower control limit** using the formula $\overline{\overline{X}} - A_2\overline{R}$ to get a lower control limit of -9.0

6) **Create a scatterplot**, graphing $\overline{X}$ on the vertical and time (day) along the horizontal.

7) At this point you will need to determine the difference between the UCL and the mean of the means, the LCL and the mean of the means.

8) Work to set your horizontal gird lines so that they represent these three lines (UCL, LCL and $\overline{X}$)

9) Once you are satisfied that you accurately display the information required for the control chart, label the horizontal lines on your graph. This can be done using a textbox.

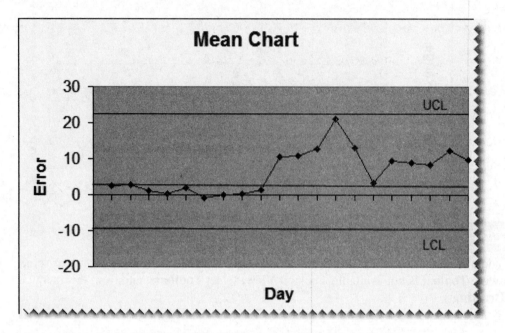

10) To add a textbox click on **Textbox** on the **Drawing Toolbar** at the bottom of your screen. (If the Drawing Toolbar is not available, click on **View,** select **Toolbars**, and from this menu, click in front of **Drawing**.)

TO PRACTICE THESE SKILLS

You can apply the skills learned in this section by working on exercises 9 and 10 from Section 14-2 Basic Skills and Concepts in your textbook..

SECTION 14-3: CONTROL CHARTS FOR ATTRIBUTES

In section 14-2 we worked with quantitative data. In this section we will work with qualitative data. We will again be selecting samples of size n at regular time intervals and plot points in a sequential graph with a centerline and control limits.

We will use the information presented in the **Defective Aircraft Altimeters** example found in section 14 – 3 of your textbook.

1) **Begin by entering the** number of defects into a new Excel worksheet.

2) **To find** $\overline{p}$

     a. **Find the sum** of the values in column A

     b. **Determine** the total number of altimeters sampled, in this case 1200.

    c. **Divide** the sum by the total number of altimeters sampled. $\overline{p} = 0.0275$

3) **Determine** $\overline{q} = 1 - \overline{p} = 1 - 0.0275 = 0.9725$

4) **Determine the upper and lower control limit** using the formula $\overline{p} \pm 3\sqrt{\dfrac{\overline{pq}}{n}}$. The **upper control limit** is 0.0766 and the **lower control limit** is $-\,0.0216$.

5) **Determine the proportion** of the sample that each defect represents. Divide each defect value by the 100, the number of altimeters in a batch.

6) **Create a scatterplot**, graphing the proportions on the vertical and number of the sample along the horizontal.

7) Work to set your horizontal gird lines so that they represent these three lines (UCL, LCL and $\overline{p}$)

8) Once you are satisfied that you accurately display the information required for the control chart, label the horizontal lines on your graph. This can be done using a textbox.

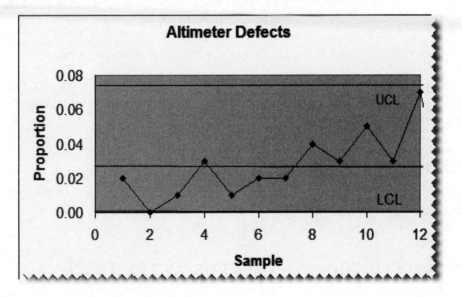

TO PRACTICE THESE SKILLS

You can apply the technology skills learned in this section by completing exercises 9, 10 and 11 from Section 14-3 Basic Skills and Concepts in your textbook.